Materials Science: Advanced Concepts

Materials Science: Advanced Concepts

Edited by **Lily Chen**

New York

Published by NY Research Press,
23 West, 55th Street, Suite 816,
New York, NY 10019, USA
www.nyresearchpress.com

Materials Science: Advanced Concepts
Edited by Lily Chen

International Standard Book Number: 978-1-63238-314-3 (Hardback)

Printed in the United States of America.

Contents

Preface

Today, materials science is an emerging field of science. It is growing internationally, which is evident by the international and intercontinental collaborations and exchanges. This book aims at providing a comprehensive coverage of selected advanced topics in materials science. This book will serve the needs of researchers and practicing scientists and will prove helpful to students of chemical engineering.

Various studies have approached the subject by analyzing it with a single perspective, but the present book provides diverse methodologies and techniques to address this field. This book contains theories and applications needed for understanding the subject from different perspectives. The aim is to keep the readers informed about the progresses in the field; therefore, the contributions were carefully examined to compile novel researches by specialists from across the globe.

Indeed, the job of the editor is the most crucial and challenging in compiling all chapters into a single book. In the end, I would extend my sincere thanks to the chapter authors for their profound work. I am also thankful for the support provided by my family and colleagues during the compilation of this book.

Editor

Materials Synthesis

Conducting Polymers / Layered Double Hydroxides Intercalated Nanocomposites

Jairo Tronto, Ana Cláudia Bordonal, Zeki Naal and
João Barros Valim

Additional information is available at the end of the chapter

1. Introduction

Layered nanocomposites represent a special class of multifunctional materials that has received a lot of attention over the last years [1-6]. The specific architecture of these composites promotes a synergistic effect between the organic and inorganic parts, generating compounds with different chemical or physical properties as compared with the isolated components. These composites not only represent a creative alternative to the search for new materials, but also allow the development of innovative industrial applications. The potential uses of layered nanocomposites include intelligent membranes and separation devices, photovoltaic devices, fuel cell components, new catalysts, photocatalysts, chemical and biochemical sensors, smart microelectronic devices, micro-optic devices, new cosmetics, sustained release of active molecules, and special materials combining ceramics and polymers, among others [7-18].

A great variety of layered nanocomposites can be prepared from the combination between polymers and layered inorganic solids [1-3]. Compared with the unmodified polymers, the resulting materials present dramatic improvement in properties such as rigidity, chemical and mechanical resistance, density, impermeability to gases, thermal stability, and electrical and thermal conductivity, as well as high degree of optical transparency.

The first successful development concerning the combination of layered inorganic solids with polymers was achieved by researchers from Toyota®, who aimed at structural applications of the nanocomposites in vehicles. These researches prepared nanocomposites by combining nylon-6 and montmorillonite (clay) using the *in situ* polymerization method [20-22]. Research conducted over the past 10 years has shown that nanocomposites containing only a small amount of inorganic silicate (2% volume), exhibit twofold larger elastic modulus and strength without sacrificing resistance to impact. Other automobile companies began to employ this type of material in their vehicles and intensified research in this area [1-3,21].

The excellent gas barrier and vapor transmission properties of these hybrid nanocomposites have led to their application mainly in food industry, more specifically in food and drink packaging. Incorporation of layered silicate nanoparticles into polymeric matrices creates a labyrinth within the structure, which physically retards the passage of gas molecules [22]. These materials can also be used to coat storage tanks in ships and lines of cryogenic fuels in aerospace systems. Compared with the unmodified polymer, nanocomposites delay fire propagation and enhance thermal stability. In contrast to the amount of additives used in traditional fireproof polymers (60%), these nanocomposites contain low layered inorganic solid loading, typically 2-5 wt%. This is due to the formation of an insulating surface layer that not only slows degradation of the polymer, but also decreases its calorific capacity [1,2]. The decomposition temperature of these nanocomposites can be increased to 100 °C, which extends the use of these materials at ambient temperatures, as in the case of automobile engines.

With respect to environmental applications, layered inorganic solids combined with biode-gradable polymers have been employed as reinforcing agents. These materials, called "green" nanocomposites, are an attractive alternative for the replacement of petroleum derivatives in the production of plastics.

Depending on the nature of the components and on the preparation method, two main types of nanocomposites can be obtained from the association of layered compound with polymers, as shown in Figure 1:

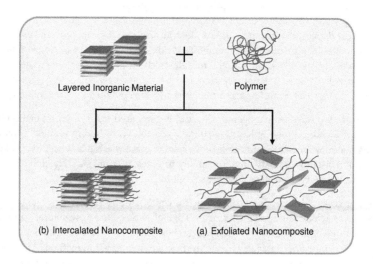

Figure 1. Schematic representation of the different types of composites produced from the interaction between layered compounds and polymers: (a) Intercalated Nanocomposite; (b) Exfoliated Nanocomposite.

- Intercalated Nanocomposite: The polymer is intercalated between the inorganic layers, producing a nanocomposite consisting of polymeric chains and alternating inorganic layers. Intercalation of the polymer often results in increased interlayer spacing; *i.e*, larger distance between two adjacent inorganic layers. (Figure 1a).

- Exfoliated Nanocomposite: The material presents no ordering along the stacking axis of the layer, or the spacing between the inorganic layers is greater than 8 nm. (Figure 1b).

- In addition to the well-defined structures cited above, a third intermediate type of structure can be found, in which the material presents characteristics of intercalation and exfoliation. In this case, there is broadening of the X ray diffraction peaks.

- Several strategies have been utilized for the preparation of organic-inorganic hybrid materials containing layered inorganic solids and polymer [1,2]:

- Exfoliation-adsorption: The layered compound is exfoliated using a solvent in which the polymer is soluble. In some layered compounds there are weak interaction forces between the layers, which can thus be easily exfoliated in appropriate solvents. The polymer may then adsorb onto the exfoliated layers which, after evaporation of the solvent, can be stacked again. As a result, the polymer is intercalated, and an ordered multilayer structure is obtained.

- *In situ* intercalative polymerization: The layered compound undergoes swelling (interlayer expansion) in a solution containing the monomer. The polymer is formed in the interlayer region. The polymerization reaction can be performed by heat or radiation treatment, using an organic initiator or a fixed catalyst.

- Melted polymer intercalation: The layered compound is mixed with the polymer matrix in the melting phase. If the layered surfaces are sufficiently compatible with the selected organic polymer under these conditions, the latter penetrates into the interlayer space, generating an intercalated or exfoliated nanocomposite. This technique does not require any solvent.

- Template Synthesis: This method can only be used for water-soluble polymers. The layered compound is formed *in situ* in an aqueous solution containing the target polymer on the basis of self-assembly forces, the polymer aids nucleation and growth of inorganic layers. As a result, the polymers are retained between the layers.

Among the inorganic solids used in the preparation of layered nanocomposites, one promising class of material is the Layered Double Hydroxides (LDHs), which have been added to polymers for the synthesis of LDH/polymers nanocomposites [23-26]. LDHs can be structurally described as the stacking of positively charged layers intercalated with hydrated anions [27]. In order to better understand the structure of the LDH, it is appropriate to start from the structure of brucite. In this $Mg(OH)_2$ structure, the magnesium cations are located in the center of octahedra, with hydroxyl groups positioned at their vertices. These octahedra share edges, forming neutral planar layers that are held together by hydrogen bonds. In this type of structure, the isomorphic replacement of bivalent cations with trivalent ones creates a positive residual charge in the layers. For charge balance to be reached in the system, anions should be present between the layers. Together with water molecules, the anions promote stacking of the layers, which culminates in the layered double hydroxide structure displaying a poorly ordered interlayer domain. Not only hydrogen bonding but also electrostatic attraction between the positively charged layers and the interlayer anions hold the layers together in LDHs. A schematic representation of the LDH structure is given in Figure 2.

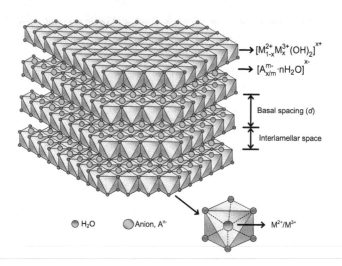

Figure 2. Schematic representation of the LDH structure.

The inorganic layers of the LDH can be stacked according to two different symmetries, resulting in rhombohedral or hexagonal unit cells of the hexagonal system. Most of the synthetic LDH belong to the hexagonal system. Only LDH with M(II)/M(III) ratio equal to 1 are orthorhombic. For the rhombohedral unit cell, the parameter c is three times the basal spacing, space group R3m. In the case of the hexagonal unit cell, the parameter c is twice the basal spacing, space group P63mmc. The notations 3R and 2H refer to unit cells as rhombohedral or hexagonal, respectively.

In the structure of LDH, the interlayer domain comprises the region between adjacent inorganic layers. This region is composed of randomly distributed anions and water molecules. Powder X ray diffraction (PXRD) and EXAFS studies, performed by Rousselet et al. showed the highly disordered nature of this region [28]. Besides being found in the interlayer domain, where they hydrate the intercalated anions, the water molecules can also be adsorbed between the crystallites. The water molecules onto the surface of the micro crystallites surface are called extrinsic water molecules, whereas those that are located in the interlayer domain are designated intrinsic water molecules. The global hydration status of the LDH is the addition of both terms, intrinsic hydration and extrinsic hydration. Many researchers consider the interlayer domain of LDH a quasi-liquid state, which gives high mobility to the interlayer anions.

A wide variety of anions can be intercalated into the LDH; for example, organic anions, inorganics and organic-inorganics, and polymers. The intercalation of more than one type of anion in the interlayer domain is an extremely rare phenomenon. Usually, the presence of two or more kinds of anions during the synthesis generates a competition between these anions, and the one with greater tendency to stabilize the system and/or that is present in larger amount will be intercalated. Using PXRD and *in situ* X ray energy dispersion spectroscopy techniques, Fogg et al. reported the existence of a second intermediate stage due to co-intercalation of Cl⁻ ions and succinate in LiAl-LDH [29]. Pisson et al. studied the exchange of Cl⁻ anions with succinate and tartarate anions in LDH of the system [Zn₂Al-Cl], [Zn₂Cr-Cl], and [Cu₂Cr-Cl]. The exchange reaction was monitored *in situ* by the X ray diffraction and X ray energy dispersion spectroscopy techniques. The analyses revealed the formation of a second intermediate stage in all the materials, caused by co-intercalation of organic anions and chloride ions [30]. Kaneyoshi and Jones demonstrated that terephthalate anions can adopt two different orientations in relation to the inorganic layers when they are intercalated into Mg-Al-LDH. The longer molecular axis is either perpendicular or parallel to the plane of the layers. These two orientations are known as interstratified intermediated phases. The occurrence of these two orientations of intercalated terephthalate anions was supported by the appearance of a third basal spacing, attributed to the contribution of two different orientations of anions in the interlayer domain.

A large number of natural and synthetic LDHs containing various metal cations have been studied. In order to form the LDH, the metal cations that will be part of the inorganic layer must present octahedral coordination and ion radius in the range of 0.50 to 0.74 Å. By varying the metal cations, the proportion among them, and the interlayer anion, a large variety of LDH can be prepared. Countless cations can be part of this structure: Mg^{2+}, Al^{3+}, most of the cations of the first transition period, Cd^{2+}, Ga^{3+}, and La^{3+}, among others [27]. In addition LDH displaying more than one bivalent and/or trivalent cation can be synthesized, which further expands the compositional possibilities.

The ratio between the metal cations M(II)/M(III) is also very important, because a change in this ratio between modifies the charge density in the layers, since the charge is generated from the isomorphic substitution of bivalent cations with trivalent ones in the structure of the inorganic layers [27,32]. There is controversy over the values that the x parameter in the general formula of the LDH can assume during the synthesis of these materials. According to de Roy et al., the x value should lie between 0.14 and 0.50, for the formation of an LDH where the M(II)/M(III) ratio can vary between 1 and 6 [33]. For Cavani et al., the x value must fall between 0.20 and 0.34, with the M(II)/M(III) ratio ranging between 2 and 4.37. However, some researchers have reported the synthesis of LDH with different M(II) to M(III) ratios from those mentioned above [27].

As described earlier, the interlayer domain consists of water molecules and anions, mainly. Practically, there is no limitation to the nature of anions that can compensate the residual positive charge of the LDH layers. However, obtaining pure and crystalline materials is not an easy task. Generally, simple inorganic anions with higher charge/radius ratio have greater tendency for intercalation. This is because these anions interact more strongly with the inorganic layers from an electrostatic viewpoint. For the intercalation of organic anions, especially anionic polymers, factors such as the size and geometry of the anion, the interaction between them, and the ratio between size and charge must be taken into account. Some interlayer anions are more mobile, which gives the resulting materials good exchange properties.

Several factors must be borne in mind when planning the synthesis of LDH. For instance, the degree of substitution of M(II) with M(III) cations, the nature of the cation, the nature of the interlayer anion, the pH of the synthesis and, in some cases, the controlled atmosphere. Furthermore, to obtain materials with good crystallinity, the concentrations of the solutions, the rate of the addition of the solutions, the stirring rate, the final pH of the suspension (for variable pH methods), the pH during the addition (for constant pH method), and the temperature of the mixture (typically performed at room temperature) must be controlled. There are a number of methods that can be used for the synthesis of LDH. They can be divided into two categories:

i.   Direct synthesis methods: salt-base method or co-precipitation (at variable pH or at constant pH), salt-oxide method, hydrothermal synthesis, induced hydrolysis, sol-gel method, and electrochemical preparation [27,33-37].

ii.   Indirect synthesis methods: simple anionic exchange method, anionic exchange by regeneration of the calcined material, and anion exchange using double phase, with formation of a salt between the surfactants [33,38,39].

Among the most extensively investigated conducting polymers are polyacetylene; poly-heterocyclic five-membered compounds like polypyrrole, polythiophene, and polyfuran; and polyaromatics such as polyaniline and poly (p-phenylene).The structures and respective electrical conductivity values of some conductive polymers are summarized in Table 1 [40].

Name	Structural Formula	Conductivity / $S \cdot cm^{-1}$
Polyacetylene		10^3 a 10^6
Polyaniline		10 a 10^3
Polypyrrole		600
Poly(p-phenylene)		500
Polythiophene		200

Table 1. Structure and electrical conductivity values of some conductive polymers [40].

Below, we will outline some classes of conducting polymers:

a. Polyaniline:

Polyanilines are widely studied because of their low cost, good stability in the presence of oxygen and water, and interesting redox properties. In 1835, polyaniline was first synthesized as "black aniline", a term used for the product obtained by oxidation of aniline [41]. Some years later, Fritzche analyzed the products obtained by chemical oxidation of this aromatic amine [42]. In 1862, Letheby found that the anodic oxidation of aniline in a platinum electrode, in an aqueous solution of sulfuric acid, formed a dark brown precipitate [43]. The polyaniline chain consists of units present in two main forms: (i) the fully reduced form, which contains only aromatic rings and nitrogen atoms of the amine function, shown in Figure 3a, and (ii) a completely oxidized form displaying iminic nitrogen atoms, quinonics rings, and aromatic rings, as represented in Figure 3b.

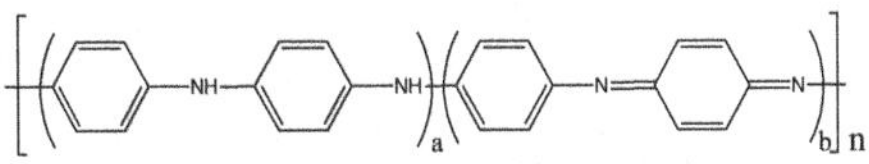

Figure 3. Representation of the general structure of the polyaniline base form: (a) reduced species (b) oxidized species.

Gospodinova and Terlemezyan examined the oxidation state of polyaniline constituents. The principal oxidation states of polyaniline are presented in Table 2 [44]. The emeraldine salt is the structural form of polyaniline that affords higher conductivity values. Polyaniline can be doped by protonation, with no change in the number of electrons in the polymer chain.

Oxidation state	Structucture	Color	Characteristic
Leucoemeraldine	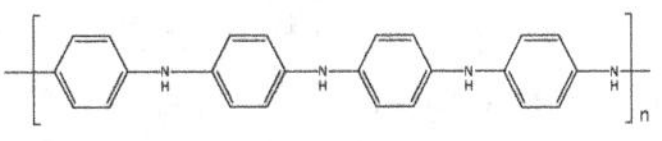	Yellow 310	Insulating, completely reduced
Emeraldine salt	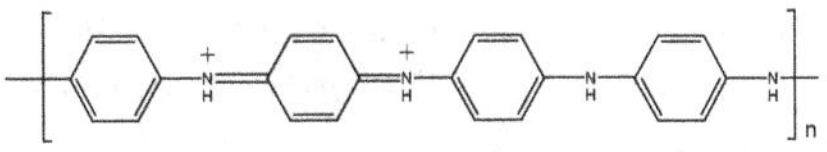	Green 320, 420, 800	Conductive, half-oxidized
Emeraldine base	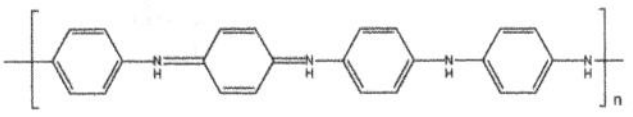	Blue 320, 620	Insulating, half-oxidized
Pernigraniline	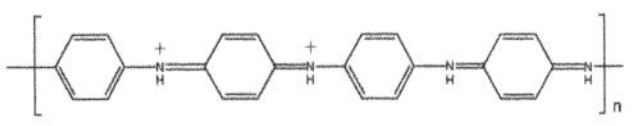	Purple 320, 530	Insulating, completely oxidized

* The numerical values refer to the wavelength (in nanometers) where absorption is maximum.

Table 2. Most important oxidation states of polyaniline [44].*

b. Polypyrrole:

The first report about pyrrole was published by Runge in 1834. This author observed a red component in coal tar and bone oil. The compound isolated and purified from this component was named of pyrrole [45]. The structural formula of pyrrole was established in 1870. In the late 19th century, the interest in pyrrole and its derivatives was aroused, following the discovery that this molecule was part of some porphyrins found in biological systems, such as chlorophyll. Approximately 100 years after the first report on the discovery of pyrrole, *i.e,* in 1970, interest in these materials increased again due to the possibility of preparing conducting polypyrroles [46].

Pyrrole is a five-membered cyclic compound (heterocyclic) containing 6 π electrons. Additionally, pyrrole has an sp^2 nitrogen, and its three σ bonds are located in the plane of the ring. The excess electron, the conjugation of the double bonds, and their ability to relocate are the structural characteristics underlying the charge conduction properties presented by the pyrrole polymer products.

c. Polythiophene:

In 1882, Meyer discovered thiophene [47]. At that time, his studies revealed that this compound, which was isolated from benzene impurities, was a new aromatic system. Thiophene is not a component of animal metabolism, but some thiophene derivatives can be found in plants. Thiophene derivatives are widely employed in many types of chemical industry, including the pharmaceutical, veterinary, polymers, and agrochemicals industries.

Thiophene is a compound analogous to pyrrole. Instead of the nitrogen heteroatom, it contains a sulfur atom with sp^2 hybridization. The sp^2 orbital, which is perpendicular to the π electron system, has an unshared electron pair. The p-orbital of sulfur donates two electrons to the π system. Polythiophene derivatives have been extensively studied, probably because most of them are soluble in organic solvents, which facilitates processing of the material.

The electrical conductivity of a solid is the result of the number of charge carriers (electrons / holes) and their mobility. Conducting polymers have a large number of charge carriers with low mobility, which is mainly caused by the large number of structural defects such as reticulation and the disordering of chains. The formation of nanocomposites by intercalation of conductive polymers into LDH can minimize the formation of reticulation defects and the disordering of polymer chains, furnishing materials with new and interesting properties. Table 3 summarizes the literature works on the synthesis and characterization of conducting polymers intercalated into LDH.

Year	Nanocomposites	Examples	Authors	Ref.
1994	LDH / polyaniline	CuCr- polyaniline-LDH CuAl- polyaniline-LDH	Challier and Slade	48
2001	LDH / aminobenzoate derivatives	LiAl-o-, p- and m-aminobenzoate-LDHs	Isupov et al.	49

2002	LDH / aniline sulfonate	CuCr- aniline sulfonate-LDH	Moujahid et al.	50
2003	LDH / aminobenzene sulfonate	CuCr- aminobenzene sulfonate-LDH	Moujahid et al.	51
2004	LDH / 2-thiophenecarboxylate	ZnAl-2-thiophenecarboxylate-LDH ZnCr-2-thiophenecarboxylate-LDH	Tronto et al.	52
2005	LDH / aniline sulfonate derivatives	CuCr-o- and m-aminobenzenesulfonate, 3-amino-4-methoxybenzenesulfonate, 3-aniline-1-propane sulfonate, and 4-aniline-1-butane sulfonate-LDHs.	Moujahid et al.	53
2006	LDH / pyrrol derivatives	ZnAl-4-(1H-pyrrol-1-yl)benzoate-LDH ZnCr-4-(1H-pyrrol-1-yl)benzoate-LDH ZnAl-3-(pyrrol-1-yl)-propanoate-LDH ZnAl-7-(pyrrol-1-yl)-heptanoate-LDH	Tronto et al.	54 55 56
2006	LDH / aminobenzoate derivatives	MgAl-aminobenzoate-LDH NiAl-aminobenzoate-LDH	Tian et al.	57 58
2006	LDH / aniline sulfonic	NiAl-aniline sulfonic -LDH	Wei et al.	59

Table 3. Some examples of Nanocomposites consisting of LDH/conductive polymers.

Challier and Slade reported the synthesis and characterization of layered nanocomposites of CuCr and CuAl-LDHs intercalated with polyaniline [48]. The oxidizing host matrices were prepared by the coprecipitation method, with the intercalation of terephthalate anions into CuCr-LDH and hexacyanoferrate(II) anions into CuAl-LDH. Then, the LDH precursors were submitted to an anion exchange reaction with a solution of pure aniline under reflux, for 24 h. The X ray diffractograms showed that the materials submitted to reaction with aniline exhibited basal spacings of 13.3 Å and 13.5 Å for CuCr-LDH and CuAl-LDH, respectively. This result was consistent with the intercalation of aniline molecules containing aromatic rings oriented perpendicular to the plane of the layers. FTIR analyses evidenced polymerization of the aniline molecules, since the absorption spectra displayed bands typical of the emeraldine form. According to the authors, the oxidant character of Cu^{2+}, present in layered structure of the inorganic host, helped induce oxidative polymerization of the aniline intercalated in the interlayer galleries.

Isupov et al. described the intercalation of o-, p-, and m-aminobenzoate anions into LiAl-LDH [49]. The incorporation of aminobenzoate anions in the host matrices was conducted by anion exchange, from an LiAl-LDH intercalated with chloride anions. The basal spacings obtained from the X ray diffractograms indicated that the anion exchange reaction was effective, with incorporation of aminobenzoate anions in the interlayer domain. To carry out the in situ polymerization, samples of the nanocomposites were submitted to a heat-treatment at 90 °C for 100 h, with 75% relative air humidity. For the LiAl-LDH intercalated with m-aminobenzoate anions, the formation of a polyconjugated system was confirmed by ESR spectra performed in vacuum at 77 K and 300 K. The spectra displayed a broad isotropic signal between 7.3 and 7.5 G, with g = 2.000 and line with Gaussian shape. Heating of the nanocomposite in air intensified the ESR signal. Formation of the polyconjugated system was also corroborated by FTIR and Raman spectroscopies. A comparison of the FTIR spectrum of LiAl-LDH intercalated with m-aminobenzoate and its oxidation products revealed a marked decrease in the band located at 1250 cm^{-1}, which indicates a decrease in the number of amino groups, -NH$_2$.

Moujahid et al. reported the intercalation of the m- and o-aminebenzeno sulfonate anion, 3-amine-4-methoxybenzene sulfonate, 3-aniline-1-propane sulfonate and 4-aniline-1-butane sulfonate into Cu$_2$Cr-LDHs [50,51,53]. These authors discussed the arrangement of intercalated molecules and their subsequent dimerization and/or in $situ$ polymerization. The authors incorporated these inorganic anions between the layers was by the direct precipitation method at constant pH. After the synthesis, the resulting materials were submitted to heat treatment at different temperatures, under air atmosphere. The interlayer distances of they synthesize nanocomposites were consistent with the presence of a bilayer of guest molecules in the interlayer space. The heat treatment performed at 350 K culminated in a contraction in the basal spacing of the nanocomposites, except for the material intercalated with the o-amine-benzene sulfonate anions. This contraction was associated with reorientation of the intercalated molecules and/or with an in $situ$ polymerization. At temperatures above 350 K up to approximately 450 K, there was no significant variation in basal spacing. ESR analysis evidenced in $situ$ polymerization. The profiles of the ESR spectra changed with increasing temperature. The value of the signal g (g = 2.0034 ± 0.0004) was typical of the formation of organic radicals and/or conduction electron. For the nanocomposite intercalated with the m-aminobenzene sulfonate anion, the ESR and CV analysis showed that the in $situ$ polymerization must occur with a syndiotactic arrangement. For the nanocomposite intercalated with the o-aminobenzene sulfonate anion, the ESR studies indicated a very weak response of spin carriers for these materials. The electrochemical characterization did not show the presence of reversible redox processes. These results, together with the constancy of the basal spacing value obtained up to a temperature of 450 K, suggested that in $situ$ polymerization was not favored when the monomer had the amino group located at the $ortho$ position relative to the sulfonate group. For the 3-amine-4-methoxybenzene sulfonate anion, the presence of methoxy group in the $para$ position relative to the sulfonate group made the polymerization process difficult. The ESR and CV data for this nanocomposite indicated formation of a dimer. For the nanocomposites synthesized with 3-aniline-1-propane sulfonate and 4-aniline-1-butane sulfonate, the heat-treatment at 473 K prompted an increase in g (g = 2.0034 ± 0.0004), which is associated with the generation of organic radicals and/or conduction electrons.

Tronto et al. described the synthesis, characterization, and electrochemical investigation of 2-thiophenecarboxylate intercalated into ZnAl-LDH and ZnCr-LDH [52]. The materials were synthesized by the coprecipitation method at constant pH, followed by hydrothermal treatment at 65 °C for 72 h. The LDH were analyzed by PXRD, FT-IR, ^{13}C CP-MAS, TEM and CV. The basal spacing was about 15.3 Å for all the LDH which suggested the formation of bilayers of anions intercalated between the inorganic sheets. In this configuration, the 2-thiophenecarboxylate anions would be in a position in which their longer axes would lie perpendicular to the plane of the inorganic layers. Besides the phase with basal spacing of 15.3 Å, another phase with basal spacing of 7.58 Å was also detected in the diffractograms. This value was similar to some values reported for the intercalation of CO_3^{2-} anions into ZnAl-LDH and ZnCr-LDH. However, the qualitative analysis and ^{13}C CP-MAS did not confirm the presence of carbonate anions, as contaminant in the LDH. Thus, the results indicated that for this second phase, 2-thiophenecarboxylate anions were intercalated with their longer axes parallel to the plane of the inorganic layer. ^{13}C CP-MAS data further suggested that, during the synthesis, the 2-thiophenecarboxylate anions lost an acid hydrogen which led to formation of the dimer.

Tronto et al. conducted a study on the *in situ* polymerization of pyrrole derivatives, 4-(1H-pyrrol-1-yl)benzoate, 3-(Pyrrol-1-yl)-propanoate, and 7-(pyrrol-1-yl)-heptanoate, intercalated into LDH [54-56]. The materials were synthesized by co-precipitation at constant pH, followed of hydrothermal treatment for 72 h. The final LDH were characterized by X ray diffraction, ^{13}C CP-MAS NMR, TGA, and ESR. The basal spacing value coincided with the formation of bilayers of intercalated monomers. ^{13}C CP-MAS NMR and ESR analyses showed the formation of a polyconjugated system with polymerization of the monomers intercalated in the LDH during the coprecipitation and/or hydrothermal treatment processes. This result reinforced the authors assumption that the connectivity between the monomers occurred spontaneously during the synthesis, with generation of oligomers and/or syndiotactic polymers intercalated between the LDH layers. At room temperature, the ESR spectrum displayed a signal typical of the hyperfine structure (hfs). The presence of hfs suggested the existence of a proper regulatory environment for the free electrons. These electrons would be present in an organic "backbone" of small size. Thermal analysis of these materials revealed that the inorganic host matrix provided the intercalated polymers with thermal protection, because the thermal decomposition reactions happened at higher temperatures compared with the pure polymers.

Tian and cols. investigated the oxidative polymerization of m-$NH_2C_6H_4SO_3^-$ anions intercalated into NiAl-LDH using ammonium persulfate as the oxidizing agent [57]. The amount of oxidizing agent required for controlled polymerization of the intercalated monomers was systematically evaluated. The materials were characterized by PXRD, UV-Vis spectroscopy, FT-IR spectroscopy, and XPS determination. PXRD and elemental analysis data showed the co-intercalation of nitrate anions, originating from the LDH precursor, and m-$NH_2C_6H_4SO_3^-$ anions. UV-Vis results evidenced polymerization of the intercalated m-$NH_2C_6H_4SO_3^-$ anions, with the formation of small chains. The intercalated polyaniline sulfonate was present in different oxidation states and at different protonation levels, depending upon the amount of oxidizing agent that was added.

Tian and cols. also performed the *in situ* oxidative polymerization of m-$NH_2C_6H_4SO_3^-$ anions intercalated into MgAl-LDH [58]. The monomers were incorporated into the LDH via an exchange reaction using the precursor $[MgAl(OH)_6](NO_3) \cdot nH_2O$. The nitrate anions remaing from the exchange reaction and co-intercalated with the m-$NH_2C_6H_4SO_3^-$ anions were utilized as oxidizing agent for the oxidative polymerization of the intercalated monomers. The resulting materials were analyzed by DTA-TG-DSC as well as UV-Vis and HT-XRD spectros-copies. In the temperature range 300-350 °C, the UV-Vis analysis confirmed the reduction of nitrate and polymerization of aniline.

Wei et al. reported the oxidative polymerization of m-$NH_2C_6H_4SO_3^-$ anions in NiAl-LDH, using intercalated nitrate anions as the oxidizing agent [59]. The LDH interlayer space was used as a "nanoreactor" for the *in situ* polymerization of the intercalated monomer. Polymerization of the monomers was acomplished by heat treatment under nitrogen atmosphere. The interlayer polymerization was monitored by thermogravimetric analysis coupled with differential thermal analysis and mass spectrometry (TGA-DTA-MS), UV-Vis spectroscopy, X ray absorption near edge (XANES), (HT-XRD) and FTIR spectroscopy. Polymerization of the monomer was observed at a temperature of 300°C.

2. Synthetic strategies for the preparation of conducting polymers / layered double hydroxides intercalated nanocomposites

The synthesis of intercalated nanocomposites of LDH/conductive polymers can be carried out using different strategies. The main ones are [23]:

1. Intercalation of monomer molecules between the LDH layers, with subsequent *in situ* polymerization. Intercalation of the monomer can occur by direct or indirect methods. The intercalation of monomer molecules with subsequent *in situ* polymerization, is widely used in the preparation of various LDH/conductive polymers. The resulting nanocomposites generally exhibit good structural organization and phase purity. This process is limited by two factors:

    i.   the distance between the monomers when they are strongly linked, or grafted, to the structure of the inorganic layers. When the monomers are strongly bound to the layers, their flexibility (freedom of movement within the interlayer) is limited, so the proximity between them should be sufficient for the polymerization reaction to occur. High charge densities in the layers may shorten distance between the intercalated monomers. Functionalized monomers with long chain aliphatic groups also provide greater flexibility.

    ii.   the polymerization conditions (temperature, pH, or redox reaction), which should be selected so as not to affect the layered structure of the resulting materials.

Indirect methods may also be employed for the intercalation of monomers. These methods are often utilized when the chemical nature of the interlayer space and guest species are not compatible. Such methods require the preparation of an LDH precursor intercalated with a molecule that can be easily exchanged. This LDH precursor is then placed in contact with the monomer of interest, which will replace the previously intercalated anion. To obtain the LDH/polymer, it is necessary to carry out the *in situ* polymerization reaction after the exchange with the monomer.

2. Direct intercalation of polymer molecules with low molecular weight between the LDH layers or intercalation polymers with high molecular weight by indirect methods. The incorporation of the polymer between the LDH layers, can be performed by direct method by using the direct co-precipitation reaction, nanocomposites containing polymers that have an anionic group; for example, carboxylate or sulfonate groups, can be produced

during growth of the inorganic crystal. This preparation strategy usually yields nano-composites with low structural organization. The crystallinity of these materials can be improved by hydrothermal treatment. The indirect method requires the presence of the LDH precursor, usually containing chloride anions. This LDH precursor is placed into exchange reaction using suitable solvents in the presence of the polymer of interest.

3. Intercalation of LDH via exfoliation, when a colloidal system is formed between the LDH and an appropriate solvent, for exfoliation of the layers. Restacking of the layers in the presence of a solution containing the target monomer or polymer culminates in their intercalation by restacking of the structure of the layer. When the monomers are interca-lated, a subsequent *in situ* polymerization is required for attainment of the intercalated nanocomposite LDH/polymer. This strategy is usually employed when the polymer has high molecular weight, which makes their diffusion between the LDH layers difficult. Due to its high charge density, the LDH does not have a natural tendency to exfoliation. To achieve delamination of these materials, it is necessary to reduce the electrostatic interaction between the layers. This can be done with intercalation of spacer anions, such as, dodecylsulfonate and dodecylbenzenesulfonate. Exfoliation is then obtained by placing the organically modified LDH in a solution containing a polar solvent. Addition of polymer to the solution containing the exfoliated material results in the formation of an intercalated and/or exfoliated precipitate. In some cases, the nanocomposite is only generated upon evaporation of the solvent.

In addition to the strategies described above, immobilization of the polymer between the LDH inorganic sheets can also be attained by regeneration of the layered structure using the "*memory effect*" exhibited by some LDH. In this case, a previously prepared LDH, normally MgAl-CO$_3$, is firstly calcined at an adequate temperature, for elimination of the interlayer anion. The calcined material, a mixed oxy-hydroxide, is then placed in contact with an aqueous solution of the polymer to be intercalated. The oxide is hydrolyzed with regeneration of the LDH structure and intercalation of the polymer. This process is accompanied by a sharp increase in the pH value. The latter can be corrected, to prevent the intercalation of hydroxyl anions. Normally, the LDH/polymers nanocomposites produced by this method do not exhibit good organization, being more suitable for the incorporation of small molecules. This method was used for the intercalation of silicates into LDH. In this case, mexinerite (an MgxAlOH-LDH, with x = 2, 3, 4) was employed as precursor for incorporation of the silicate to this end, mexinerite was previously calcined at 500 °C under air atmosphere, and then placed in contact with a solution of tetraethylorthosilicate, Si(OC$_2$H$_5$)$_4$ (TEOS). This afforded more crystalline materials than those obtained by anion exchange or direct co-precipitation, using metasilicate and ZnM-LDHs (M = Al, Cr).

An additional route for preparation of the LDH/polymer is the auxiliary solvent method. Solvents represent an important part in the swelling processes of the layered materials, since they promote separation of the layers. Schematic representation of the incorporation of polymers into layered double hydroxides is given in Figure 4.

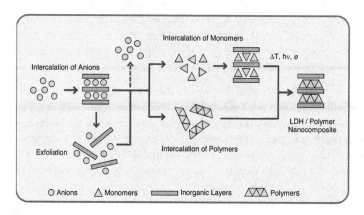

Figure 4. Schematic representation of the incorporation of polymers into layered double hydroxides. *(adapted from ref. 23)*

3. Characterization methods

This section describes the main techniques employed for the characterization of intercalated nanocomposites of conducting polymer / LDH: Powder X-ray Diffraction (PXRD), ^{13}C Cross-Polarization/Magnetic Angle Spinning (CP/MAS) NMR spectroscopy, Electron Spin Resonance (ESR) spectroscopy, Thermogravimetric Analysis (TGA), Differential Scanning Calorimetry (DSC), Fourier Transform Infrared (FTIR) spectroscopy, Ultraviolet/Visible (UV-Vis) Spectroscopy, Transmission Electron Microscopy (TEM), and Scanning Electron Microscopy (SEM).

3.1. Powder X-ray Diffraction (PXRD)

The X ray diffraction pattern (PXRD) of LDH presents basal peaks 00l related to the stacking sequence of the inorganic sheet. The peaks are not basal, said to non harmonics, are related to the sheet structure. For new LDHs, the indexing of the diffraction peaks can be accomplished by comparison with the PXRD of hydrotalcite, which exists in the database of diffraction equipment (JCPDS-ICDD, PDF database), or with a number of other LDHs described in the literature. Figure 5 brings a representative PXRD for an MgAl-CO$_3$-LDH.

The interlayer distances can be calculated from the values of 2θ, using the Bragg equation:

$$n\lambda = 2d_{hkl} \cdot sen\theta$$

where n is the diffraction order, d_{hkl} is the interlayer spacing for the peak hkl, and θ is the Bragg angle, determined by the diffraction peak. Repetition of the d value, for n = 1, 2, 3..., evidences the formation of a layered material. The interlayer spacing can be calculated by averaging the basal peaks according to the equation:

$$d=\frac{1}{n}(d_{003}+2d_{006}+...+nd_{00n})$$

The parameters a and c can be obtained according to the equation:

$$\frac{1}{(d_{hkl})^2}=\frac{4}{3}\left(\frac{h^2+hk+k^2}{a^2}\right)+\frac{l^2}{c^2}$$

where h, k, and l are the Miller indices of the corresponding peak. For a LDH with stacking sequence 3R, the c parameter c is equal to three times the basal spacing value.

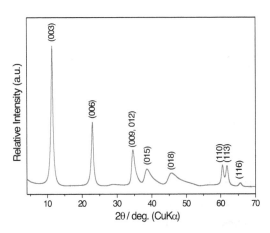

Figure 5. PXRD of synthetic Hydrotalcite.

To determine the orientation adopted by anionic species, such as monomers and polymers intercalated into LDHs, the values of interlayer spacing and/or basal spacing obtained from the PXRD data are compared with the size of anions obtained by specific computer programs, like "VASP (*Vienna Ab-initio Simulation Package*)".

When thermal treatments is performed for the *in situ* polymerization of monomers intercalated between the LDHs inorganic layers, the PXRD analysis may reveal a decrease in the value of interlayer spacing, which indicates a small contraction between adjacent layers. The presence of phases, other than the LDH can also be identified by PXRD, which is useful since thermal treatment may often generated oxides.

3.2. ¹³C Cross-Polarization/Magnetic Angle Spinning (CP/MAS) NMR spectroscopy

In situ polymerization of monomers intercalated into LDH may be monitored by ¹³C Cross-Polarization/Magnetic Angle Spinning (CP/MAS) NMR spectroscopy. This technique detects formation of bonds of the monomer-monomer type in polyconjugated systems. Assignment of the chemical shift values for the monomers can be carried out by computer simulation using specific computer programs, such as *"ACD/ChemSketch, version 4.04"*, provided by the company Advanced Chemistry Development Inc., and *"CS Chemdraw Ultra®"*, offered by the company Cambridgesoft Corporation. The values obtained by simulation can be compared with the values achieved experimentally.

Figure 6 contains an example of ^{13}C Cross-Polarization/Magnetic Angle Spinning (CP/MAS) NMR Spectroscopy analyses for the *in situ* polymerization of 4-(1H-pyrrol-1-yl)benzoate intercalated into ZnAl-LDH [54]. Assignment of the peaks to the carbons of the monomer is given in Table 4.

Notation	Assignment (Cn)	DMSO-D6 (ppm)	CP-MAS (75.4 MHz) (ppm)
C1	C1	111.3	111.2
C2	C2	118.9	120.1
C3	C3	143.0	142.3
C4	C4	118.5	115.6
C5	C5	130.9	132.9
C6	C6	127.0	122.8
C7	C7	166.6	174.9

Table 4. Assignment of the peaks to the carbons in the ^{13}C CP-MAS NMR analyses of 4-(1H-pyrrol-1-yl)benzoate.

The ^{13}C CP/MAS NMR spectra of all LDH were similar. In the Figure 6c and 6d, the peaks can be unambiguously assigned as carbons C7 (175.1 ppm), C5 (131.6 ppm), C3 (140.8 ppm), and C1 (113.7 ppm). The broad signal at 116.8 ppm can be attributed to the chemical shifts of the remaining carbons C4 and C6 of the six-membered ring. Several simulations of the ^{13}C NMR spectra suggest that one possible quaternary carbon, resulting from the polymerization of the monomer via condensation C2-C2, presents chemical shift in the range of 112.0 to 116.0 ppm. Therefore, the large signal at 116.8 ppm in spectrum of the polymer is ascribed to this quaternary carbon, coinciding with the chemical shifts of the remaining carbons C4 and C6. Together with the PXRD results, these data suggest that the production of oligomers and/or polymers occurs with the formation of bilayers of monomers in the interlayer space. In this arrangement, the carboxylate groups are directed to the layer, whereas the aromatic rings occupy the central region of the interlayer spacing. Therefore, the polymer obtained within the interlayer resembles a *"zig-zag"*, similar to the polymers of the syndiotactic type.

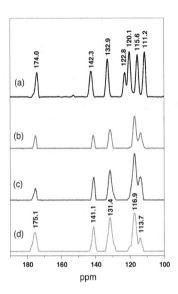

Figure 6. CP-MAS NMR spectra of: (a) pyrrolebenzoic acid; (b) ZnAl-4-(1H-pyrrol-1-yl)benzoate-LDH; (c) ZnAl-4-(1H-pyrrol-1-yl)benzoate-LDH with hydrothermal treatment; and (d) ZnAl-4-(1H-pyrrol-1-yl)benzoate-LDH after thermal treatment [54].

3.3. Electron Spin Resonance (ESR) spectroscopy

Electron Spin Resonance (ESR) spectroscopy allows for monitoring of *in situ* polymerization processes in intercalated monomers. The spectra of conducting polymers usually exhibit signs typical of the formation of polarons, with the Lorentzian profile. In these analyses, the apparatus is normally operated at 9.658 GHz, using the 1,1-diphenyl-2-picrylhydrazyl (DPPH) radical to determine the ressonance frequency (g = 2.0036 +/- 0.0002). The scan width can vary between 2000 and 4000 G, with a receiver gain of 100000.

Figure 7 illustrates ESR analyses for monitoring of the *in situ* polymerization of 3-(Pyrrol-1-yl)-propanoate monomers intercalated into ZnAl-LDH [55,56]. The spectra were recorded after heat-treatment at temperatures ranging from ambient to 180 °C for 2h. For material at room temperature, the ESR spectra display very weak signals. Thus, the spectrum was enlarged 16 times for comparison with those of the material treated at other temperatures. Typical signs can be noticed for the *"superhyperfine"* structure with 6 lines, and there is a sign characteristic of the formation of a polaron with g = 2.004 ± 0.0004. The appearance of this *"superhyperfine"* structure suggest formation of the radical (COO·). The magnetic moment of this radical should interact with the magnetic moments of the nuclei of the aluminum atoms present in the inorganic host matrix. This hypothesis considers the nuclear spin of aluminum as I = 5/2 and a number of nuclei N = 1, which generates a spectrum of 2NI + 1 = 6 lines. Due to the charge balance required for maintenance of the electroneutrality of hybrid systems there a regulating environment for the free electrons of the radicals (COO ·) within the interlayer spacing. These radicals are located near the aluminum cations, because the latter are responsible for the positive charge density of the layer. The signal at g = 2.004 ± 0.0004 attests to the formation of a polaron, *i.e.*, a polarized entity resulting from delocalization of the radical in structures with π conjugations. The increase in the delocalization of π orbitals favors the generation of polarons, so an increase signal in this upon heat-treatment indicates stronger connection between the monomers. The Lorentzian profile of the ESR spectrum of the material at room temperature is compatible with the formation of conjugated polymers. The ESR results agree with the NMR results and indicate that spontaneous partial polymerization and/or oligomerization of the 3-(Pyrrol-1-yl)-propanoate monomers takes place during coprecipitation of the nanocomposites.

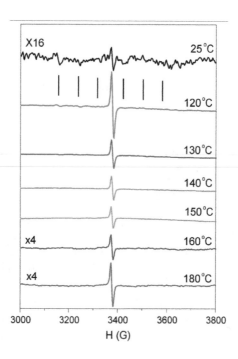

Figure 7. ESR spectra of Zn-Al-3-(Pyrrol-1-yl)-propanoate-LDH as a function of the heat-treatment temperature [55,56].

3.4. Thermogravimetric Analysis (TGA) and Differential Scanning Calorimetry (DSC)

The thermal stability of LDHs intercalated with conductive polymers as well as the amount of water, intercalated and adsorbed, in the nanocomposites can be determined by thermogravimetric analysis. The results are obtained as a curve mass decrease (%) *versus* temperature. For the LDHs, the thermal decomposition steps generally overlap, especially in the case of LDHs intercalated with organic molecules.

TGA is important in the thermal *in situ* polymerization of nanocomposites, since it is necessary to determine the temperature that should be used for polymerization of the intercalated monomers. The thermal decomposition of intercalated organic compounds takes place at higher temperatures, so it is possible to achieve greater thermal stability for conducting polymers intercalated into an inorganic host matrix (LDH).

Figure 8 displays an example of TGA/DSC analysis for the ZnAl-LDH intercalated with 3-aminobenzoate monomers and for the pure monomer.

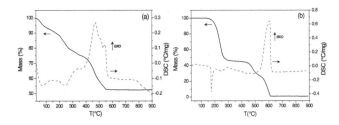

Figure 8. TGA-DSC of (a) ZnAl-3-aminobenzoate-LDH; (b) sodium 3-aminobenzoate [60].

For the nanocomposite material, Figure 8a, the early stages of thermal decomposition are associated with loss of adsorbed and intercalated water. In this temperature range, the DSC curve reveals the occurrence of endothermic processes. Dehydroxylation of the inorganic sheets and decomposition of the anion intercalated species happen concomitantly. The DSC curve also indicates the occurrence of an exothermic process during decomposition of the intercalated organic species.

3.5. Fourier Transform Infrared (FTIR) spectroscopy

FTIR analysis is carried out in KBr pellets, pressed from a mixture of 2% of the LDH samples in previously dried KBr. The spectra are recorded over a wavelength range going from 4000 to 400 cm^{-1}. FTIR spectroscopy data provide information about the functional groups and possible interactions between the organic and inorganic parts of the nanocomposites. Identification of the *in situ* polymerization of monomers intercalated into LDH by this technique is difficult because of several overlapping spectral bands.

Figure 9 contains the FTIR spectra of (a) sodium 3-aminobenzoate, (b) pure sodium poly-3-aminobenzoate, and LDH intercalated with sodium 3-aminobenzoate submitted to different treatments [61]. For the pure monomer, the bands (not shown in the figure) at 3408, 3349, and 3223 cm^{-1} are related to $\nu(NH_2)$ symmetric and anti-symmetric stretching, whereas the band at 1628 cm^{-1} is characteristic of $\delta(NH_2)$ symmetric deformation. The bands at 1560 and 1411 cm^{-1} are typical of ν(-COO$^-$) symmetric and anti-symmetric stretching, respectively. The band at 1312 cm^{-1} is due to ν(C-N) stretching. The bands at 1266 and 1115 cm^{-1} are attributed to δ(NH$_2$) symmetric and asymmetric deformations. The bands at 776 and 676 cm^{-1} are ascribed to δ(COO$^-$) out of the plane symmetric and asymmetric deformation. Concerning the polymer, the FT-IR spectrum of pure poly-3-aminobenzoate undergoes significant changes, especially in the area relative to the vibrations of the aromatic ring and the functional group NH$_2$. The bands due to (-COO$^-$) in the plane stretching at the 1700 and 1400 cm^{-1}, and (NH$_2$) out of the plane deformation at 1698, 1634, 1566, 1509 and 1441 cm^{-1} are fairly broad. The broad over-lapping bands in the region between 1300 and 1110 cm^{-1}, refer to δ(NH$_2$) symmetric and asymmetric. Alterations in the spectrum of the polymer are expected because the amine and p-methylenes groups of the 3-aminobenzoate molecules interact during polymerization. As for the heat-treated nanocomposites, there is virtually no changes in the profile of the spectra. The bands in the regions between 1700 and 1360 cm^{-1}, related to stretching of the carboxylate group and the aromatic ring are displaced and broader. The bands relative to ν(-COO$^-$) symmetric and anti-symmetric stretching can be observed in the regions near 1554 and 1384 cm^{-1}. In the region between 1300 and 1110 cm^{-1} there is a shoulder around 1303 cm^{-1} and weak at 1266 cm^{-1}, corresponding to δ(NH$_2$) symmetric and asymmetric deformation. Analysis of the bands in the regions below 1200 cm^{-1} is highly compromised because of the large overlap of bands with medium and weak intensity. The bands in the regions below 700 cm^{-1} are due to metal-oxygen-metal vibrations occurring in the inorganic host matrix.

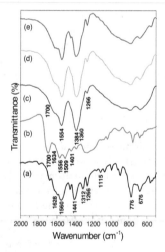

Figure 9. FTIR spectra of (a) sodium 3-aminobenzoate; (b) pure sodium poly-3-aminobenzoate; (c) MgAl-3-aminobenzoate-LDH; (d) MgAl-3-aminobenzoate-LDH with hydrothermal treatment; and (e) MgAl-3-aminobenzoate-LDH heat-treated at 160 °C [61].

3.6. Ultraviolet/Visible (UV-Vis) spectroscopy

The UV-Vis spectra are collected between 200 and 800 nm. Samples are prepared by dissolution of the material in concentrated HCl and subsequent dilution in water.

Figure 10 depicts the UV-Vis spectra pure sodium Poly-3-aminobenzoate and ZnAl-AMB-LDH with different Zn:Al molar ratios(2:1, 3:1 and 6:1) [60].

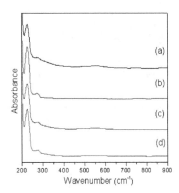

Figure 10. UV-Vis absorption spectra of the materials prepared with (a) pure sodium poly-3-aminobenzoate; (b) Zn$_2$Al-3-aminobenzoate-LDH; (c) Zn$_3$Al-3-aminobenzoate-LDH; and (d) Zn$_6$Al-3-aminobenzoate-LDH.

All the LDH display a band at about 225 nm, after polymerization, the band verified for the monomers is dislocated to lower wavelengths ~215 nm, and a band at ~ 275 nm appears. The latter band is less pronounced for Zn$_6$Al-3-AMB-LDH prepared by anion exchange in double phase, which is attributed to the n-π* transition due to the presence of non-shared electrons in the COO⁻ group. After polymerization a peak at ~ 315 nm ascribed to π-π* transition related to conjugation of rings in the polymeric chain is detected. As for the LDH, the first absorption peak intensifies ongoing from the compounds prepared with Zn/Al ratios of 2:1 and 3:1 to 6:1. In the case of the materials prepared by exchange in double phase only for the compounds with Zn/Al ratios of 2:1 and 3:1 the band intensifies. The compound with Zn/Al ratio of 6:1 has the least intense peak.

3.7. Transmission Electron Microscopy (TEM)

The best TEM images are generally achieved when LDHs are dispersed in an epoxy resin, centrifuged, and kept at 70 °C for the 72 h, for drying. After drying, the materials are cut in an ultra-microtome and transferred to hexagonal copper bars appropriated for TEM image acquisition. An alternative approach is to prepare a suspension containing ethanol and LDH. The copper grid is then immersed into the suspension and dried at ambient temperature.

Figure 11 reveals very orderly particles in which the darkest lines represent the inorganic layers and the clearest lines refer to the intercalated conductive polymers [54]. There is good pillaring of the sheets, with a large sequence of darker lines. The basal spacing value estimated from the TEM images can be compared with the one obtained PXRD analysis.

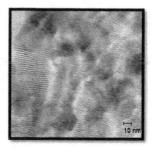

Figure 11. TEM micrographs for ZnAl-4-(1H-pyrrol-1-yl)benzoate-LDH with hydrothermal treatment.

3.8. Scanning Electron Microscopy (SEM)

The morphology of the crystallites and nanocomposite particles can also be analyzed by SEM.

For these analyses, the samples are usually supported on the sample port by powder dispersion on double-sided conductive adhesive tape. Because LDHs do not present enough conductivity for generation of good images it is necessary to cover the samples with gold before the measurements, using a sputter equipment.

Figure 12 shows the SEM images of LDH intercalated with 3-aminobenzoate. There is superposition of the sheets, with formation of aggregates on the surface of the cristallyte [60].

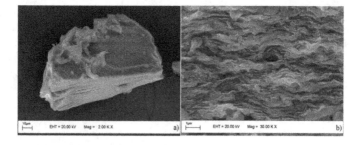

Figure 12. SEM images of ZnAl-3-aminobenzoate-LDH [60].

3.9. Cyclic Voltammetry (CV)

Cyclic Voltammetry (CV) experiments are conducted on potentiostats. The supporting electrolyte is $0.1 \ mol/dm^3 \ LiClO_4$ solution, and a conventional electrochemical cell arrangement involving three electrodes is utilized: Platinum wire as the counter electrode, as the reference electrode (Ag/AgCl/KCl(sat)), and glassy carbon, prepared by dip-coating in an aqueous suspension of monomers intercalated into LDH as the working electrode. The potential of the liquid junction is disregarded. CV experiments enable evaluation of the oxidation and reduction processes of the intercalated monomers. A typical voltammogram of ZnAl-LDH intercalated with 3-aminobenzoate anions is presented in Figure 13. The oxidation process involved in the polymerization of 3-aminobenzoate intercalated into LDH can be noticed. Moreover, Zn^{2+} oxidation can be verified.

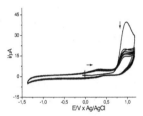

Figure 13. CV for ZnAl-3-aminobenzoate-LDH [60].

There is an irreversible oxidation peak at about 0.960V, and the amplitude of this peak diminishes upon consecutive scanning. This peak is ascribed to 3-aminobenzoate oxidation. A similar behavior has been previously observed for 2-thiophenecarboxylate anions intercalated into ZnAl-LDH.

4. Conclusion

Layered Double Hydroxides (LDHs) are materials whose layered architecture enables separation of the inorganic part (double hydroxide), and the organic portion (conductive polymer), thus culminating in a hybrid composite. The "*growth*" of conductive polymers in limited spaces, like the interlayer region of the LDHs, has been shown to be a very promising method for the improvement of the properties of conductive polymers.

On the basis of literature works, it is possible to deduce that, the guests species (monomers) are generally intercalated in a bilayer arrangement within the LDH layers. In this arrangement, the functional groups of the monomers are directed to the inorganic layer, and the aromatic rings occupy the central region of the interlayer spacing. The nature of the substituent group (aliphatic or aromatic) influences the structural organization and the *in situ* polymerization of the resulting hybrid materials.

During the synthesis, some nanocomposites undergo spontaneous polymerization, while others have to be submitted to thermal or electrochemical treatments to reach polymerization. Monomers containing substituents with aliphatic chains, tend to undergo polymerization in milder conditions, because the aliphatic chains provide small mobility of the intercalated monomers, thereby faciliting formation of polyconjugated systems. In some cases, thermal treatment may cause collapse of the layered structure, with consequent formation of oxide.

The thermogravimetric analysis data show that, compared with the pure polymer, the LDH-intercalated conducting polymer is more thermally stable. This stability is provided by the inorganic coverage offered by the LDH layers.

In the case of materials intercalated with conducting polymers, there is initial removal of one electron from the polymeric chain, *e.g*, through *p* doping. This results in the formation of an electronic state denominated polaron. Generation of the polaron can also be interpreted as π electron redistribution. Moreover, the formation of this entity is associated with distortion of the polymeric chain, which transforms the aromatic form into the quinoid form. The production of polaron may be also due the presence of electronic state located in the energy region found in the middle of gap. The quinoid structure presents smaller ionization energy and larger electronic affinity than the aromatic form. Polaron is chemically defined as a radical ion of spin = 1/2. As the concentration of polarons increases, they tend to recombine, stabilizing the structure and forming a "bipolaron". "Bipolaron" is defined as a pair of equal diamagnetics dication with spin equal to 0 and equal charges. The formation of "Bipolaron" is associated with strong distortion to the LDH net work.

Acknowledgements

This work was supported by the Brazilian agencies: Fundação de Amparo à Pesquisa do Estado de Minas Gerais (FAPEMIG), Fundação de Amparo à Pesquisa do Estado de São Paulo (FAPESP), and Conselho Nacional de Desenvolvimento Científico e Tecnológico (CNPq).

Author details

Jairo Tronto[1*], Ana Cláudia Bordonal[2], Zeki Naal[3] and João Barros Valim[2]

*Address all correspondence to: jairotronto@ufv.br

1 Universidade Federal de Viçosa - Instituto de Ciências Exatas e Tecnológicas - Campus de Rio Paranaíba - Rio Paranaíba - MG, Brazil

2 Universidade de São Paulo - Faculdade de Filosofia Ciências e Letras de Ribeirão Preto - Departamento de Química - Ribeirão Preto – SP, Brazil

3 Universidade de São Paulo - Faculdade de Ciências Farmacêuticas de Ribeirão Preto - Departamento de Física e Química - Ribeirão Preto – SP, Brazil

References

[1] Pinnavaia TJ, Beall GW. Eds. Polymer-Clay Nanocomposites; Jonh Wiley & Sons Ltd.: New York; 2000.

[2] Alexandre M, Dubois P. Polymer-Layered Silicate Nanocomposites: Preparation, Properties and Uses of a New Class of Materials. Materials Science & Engineering R-Reports 2000;289(1-2) 1-63.

[3] LeBaron PC, Pinnavaia TJ. Clay Nanolayer Reinforcement of a Silicone Elastomer. Chemistry of Materials 2001;13(10) 3760-3765.

[4] Lagaly G. Introduction: From Clay Mineral-Polymer Interactions to Clay Mineral-Polymer Nanocomposites. Applied Clay Science 1999;15(1-2) 1-9.

[5] Darder M, Aranda P, Ruiz AI, Fernandes FM, Ruiz-Hitzky E. Design and Preparation of Bionanocomposites Based on Layered Solids with Functional and Structural Properties. Materials Science and Technology 2008;24(9) 1100-1110.

[6] Podsiadlo P, Kaushik AK, Arruda EM, Waas AM, Shim BS, Xu J, Nandivada H, Pumplin BG, Lahann J, Ramamoorthy A, Kotov NA. Ultrastrong and Stiff Layered Polymer Nanocomposites. Science 2007;318(5847) 80-83.

[7] Merkel TC, Freeman BD, Spontak, RJ, He Z, Pinnau I, Meakin P, Hill AJ. Ultrapermeable, Reverse-Selective Nanocomposite Membranes. Science 2002;296(5567) 519-522.

[8] Ogara JE, Ding J, Walsh D; Waters investments ltd (WATE-Non-standard) Waters technologies corp (wate-non-standard), assignee. Hybrid inorganic-organic material used for separation devices, comprises polymerized scaffolding nanocomposite containing scaffolding functionality capable of chemically interacting with surface of another material patent WO2004105910-A2; GB2419886-A; DE112004000906-T5; US2007141325-A1; JP2007515503-W; GB2419886-B; WO2004105910-A3; JP2012042477-A WO2004105910-A2 09 Dec 2004 B01D-000/00 200505.

[9] Chen H, Yu H, Hsiao W, Chen X, Xiao W, You X; Du Pont Apollo Ltd (Dupo), assignee. Backsheet for a photovoltaic module, comprises a nanocomposite layer comprising a polymeric matrix including a polymer consisting of polyethylene terephthalate and silicate nanoparticles, polymeric layers, and adhesive layers patent US2011259415-A1; CN102280505-A US2011259415-A1 27 Oct 2011 H01L-031/0216 201172.

[10] Unnikrishnan L, Mohanty S, Nayak SK, Singh N. Synthesis and Characterization of Polysulfone/Clay Nanocomposite Membranes for Fuel Cell Application. Journal of Applied Polymer Science 2012;124(SI) E309-E318.

[11] Yuan S, Li Y, Zhang Q, Wang H. ZnO/Mg-Al Layered Double Hydroxides as Strongly Adsorptive Photocatalysts. Research on Chemical Intermediates 2009;35(6-7) 685-692.

[12] Cavani F, Clause O, Trifiro F, Vaccari A. Anionic Clays with Hydrotalcite-Like Structure as Precursors of Hydrogenation Catalysts. Advances in Catalyst Design, 1991; 186-190.

[13] Vial S, Prevot V, Leroux F, Forano C. Immobilization of Urease in ZnAl Layered Double Hydroxides by Soft Chemistry Routes. Microporous and Mesoporous Materials. 2008;107(1-2) 190-201.

[14] Hsu SL, Chang K, Shiu L, Jang G; HSU S L (HSUS-Individual) CHANG K (CHAN-Individual) UNIV NAT CHENG KUNG (UNCK), assignee. Fabricating Polybenzoxazole Clay Nanocomposite for Microelectronic Industry, by Ion Exchange Reaction of Modifying Agent and Layered Clay, and Polycondensation Reaction of Diacid Chloride and Bis(o-aminophenol) Monomer Patent US2003139513-A1; TW576855-A; US7081491-B2 US2003139513-A1 24 Jul 2003 B29C-039/14 200373.

[15] Bonifacio LS, Gordijo CR, Constantino VRL, Silva DO, Kiyohara PK, Araki K, Toma HE. Optical Changes and Writing on Hydrotalcite Supported Gold Nanoparticles. Journal of Nanoscience and Nanotechnology. 2008;8(1) 274-279.

[16] Carretero MI, Pozo M. Clay and Non-Clay Minerals in the Pharmaceutical and Cosmetic Industries Part II. Active ingredients. Applied Clay Science. 2010;47(3-4) 171-181.

[17] Cunha VRR, Ferreira AMD, Constantino VRL, Tronto J, Valim JB. Layered Double Hydroxides: Inorganic Nanoparticles for Storage and Release of Species of Biological and Therapeutic Interest. Quimica Nova. 2010;33(1) 159-171.

[18] Bernardo E, Colombo P, Hampshire S. Advanced Ceramics from a Preceramic Polymer and Nano-Fillers. Journal of the European Ceramic Society. 2009;29(5) 843-849.

[19] Okada A, Fukushima Y, Kawasumi M, Inagaki S, Usuki A, Kurauchi T, Kamigaito O, Sugiyama S. Toyota Chuo Kenkyushok, assignee. Composite with High Strength and Excellent High Temp. Properties has Layers of Silicate Mineral Homogeneously Dispersed in a Poly-Amide Matrix. patent DE3632865-A1; JP62074957-A; JP62252425-A; US4739007-A; JP7309942-A; JP7310012-A; JP96022946-B2; DE3632865-C2; JP2663113-B2; JP2724547-B2; US37385-E.

[20] Usuki A, Kojima Y, Kawasumi M, Okada A, Fukushima Y, Kurauchi T, Kamigaito O. Synthesis of Nylon 6-clay Hybrid. Journal of Materials Research. 1993;8(5) 1179-1184.

[21] Kojima Y, Usuki A, Kawasumi M, Okada A, Fukushima Y, Kurauchi T, Kamigaito O. Mechanical-Properties of Nylon 6-clay Hybrid. Journal of Materials Research. 1993;8(5) 1185-1189.

[22] Yano K, Usuki A, Okada A. Synthesis and Properties of Polyimide-Clay Hybrid Films. Journal of Polymer Science Part A-Polymer Chemistry. 1997;35(11) 2289-2294.

[23] Leroux F, Besse JP. Polymer Interleaved Layered Double Hydroxide: A New Emerging Class of Nanocomposites. Chemistry of Materials. 2001;13(10) 3507-3515.

[24] Wang G, Cai FL, Si LC, Wang ZQ, Duan X. An Approach Towards Nano-Size Crystals of Poly(acrylic acid): Polymerization Using Layered Double Hydroxides as Template. Chemistry Letters. 2005;34(1) 94-95.

[25] Wang GA, Wang CC, Chen CY. The Disorderly Exfoliated LDHs/PMMA Nanocomposite Synthesized by In Situ Bulk Polymerization. Polymer. 2005;46(14) 5065-5074.

[26] Darder M, Lopez-Blanco M, Aranda P, Leroux F, Ruiz-Hitzky E. Bio-nanocomposites based on layered double hydroxides. Chemistry of Materials. 2005;17(8) 1969-1977.

[27] Cavani F, Trifiro F, Vaccari A. Hydrotalcite-Type Anionic Clays: Preparation, Properties and Applications. Catalysis Today. 1991;11(2) 173-301.

[28] Roussel H, Briois V, Elkaim E, de Roy A, Besse JP. Cationic Order and Structure of [Zn-Cr-Cl] and [Cu-Cr-Cl] Layered Double Hydroxides: A XRD and EXAPS Study. Journal of Physical Chemistry B. 2000;104(25) 5915-5923.

[29] Fogg AM, Dunn JS, O'Hare D. Formation of Second-Stage Intermediates in Anion-Exchange Intercalation Reactions of the Layered Double Hydroxide [LiAl2(OH)(6)]Cl Center Dot H2O as Observed by Time-Resolved, In Situ X-ray Diffraction. Chemistry of Materials. 1998;10(1) 356-360.

[30] Pisson J, Taviot-Gueho C, Israeli Y, Leroux F, Munsch P, Itie JP, Briois V, Morel-Desrosiers N, Besse, JP. Staging of Organic and Inorganic Anions in Layered Double Hydroxides. Journal of Physical Chemistry B. 2003;107(35) 9243-9248.

[31] Kaneyoshi M, Jones W. Exchange of Interlayer Terephthalate Anions from a Mg-Al Layered Double Hydroxide: Formation of Intermediate Interstratified Phases. Chemical Physics Letters. 1998;296(1-2) 183-187.

[32] Morpurgo S, LoJacono M, Porta P. Copper-Zinc-Cobalt-Aluminium-Chromium Hydroxycarbonates and Mixed Oxides. Journal of Solid State Chemistry.1996;122(2) 324-332.

[33] de Roy A, Forano C, El Malki K, Besse JP. In Synthesis of Microporous Materials; Ocelli, M. L.; Robson, M. E. (ed.) Van Nostrand Reinhold: New York; 1992. p. 108-169.

[34] Lopez T, Bosch P, Ramos E, Gomez R, Novaro O, Acosta D, Figueras F. Synthesis and Characterization of Sol-Gel Hydrotalcites. Structure and Texture. Langmuir. 1996;12(1) 189-192.

[35] Reichle WT. Synthesis of Anionic Clay-Minerals (Mixed Metal-Hydroxides, Hydrotalcite). Solid State Ionics. 1986;22(1) 135-141.

[36] Taylor RM. The Rapid Formation of Crystalline Double Hydroxy Salts and Other Compounds by Controlled Hydrolysis. Clay Minerals. 1984;19(4) 591-603.

[37] Indira L, Dixit M, Kamata PV. Electrosynthesis of Layered Double Hydroxides of Nickel with Trivalent Cations. Journal of Power Sources. 1994;52(1) 93-97.

[38] Kooli F, Depege C, Ennaqadi A, de Roy A, Besse JP. Rehydration of Zn-Al Layered Double Hydroxides. Clays and Clay Minerals. 1997;45(1) 92-98.

[39] Crepaldi EL, Pavan PC, Valim JB. A New Method of Intercalation by Anion Exchange in Layered Double Hydroxides. Chemical Communications. 1999;2155-156.

[40] Kumar D, Sharma RC. Advances in Conductive Polymers. European Polymer Journal. 1998;34(8) 1053-1060.

[41] Syed AA, Dinesan MK. Polyaniline – A Novel Polymeric Material - Review. Talanta. 1991;38(8) 815-837.

[42] Fritsche, J. Ueber das Anilin, Ein Neues Zersetzungsproduct des Indigo. Journal für Praktische Chemie, 1840; 20 453-459.

[43] Letheby HJ. On the Production of a Blue Substance by the Electrolysis of Sulphate of Aniline. Journal of the Chemical Society, 1862;15 161-163.

[44] Gospodinova N, Terlemezyan L. Conducting Polymers Prepared by Oxidative Polymerization: Polyaniline. Progress in Polymer Science. 1998;23(8) 1443-1484.

[45] Runge F.F. Ueber Einige Produkte der Steinkohlendestillation. Annalen der Physik. 1834; 31 65-78.

[46] Skotheim TA, editor. Handbook of Conducting Polymers. M. Dekker: New York, 1998.

[47] Meyer V. Ueber Benzole Verschiedenen Ursprungs. Berichte der DeutschenChemischenGesellschaft.1882;15 2893-2894.

[48] Challier T, Slade RCT. Nanocomposite Materials - Polyaniline-Intercalated Layered Double Hydroxides. Journal of Materials Chemistry. 1994;4(3) 367-371.

[49] Isupov VP, Chupakhina LE, Ozerova MA, Kostrovsky VG, Poluboyarov VA. Polymerization of m-NH2C6H4COO Anions in the Intercalation Compounds of Aluminium Hydroxide [LiAl2(OH)(6)][m-NH2C6H4COO] center dot nH(2)O. Solid State Ionics. 2001;141(SI) 231-236.

[50] Moujahid EM, Dubois M, Besse JP, Leroux F. Role of Atmospheric Oxygen for the Polymerization of Interleaved Aniline Sulfonic Acid in LDH. Chemistry of Materials. 2002;14(9) 3799-3807.

[51] Moujahid EM, Leroux F, Dubois M, Besse JP. In Situ Polymerization of Monomers in Layered Double Hydroxides. Comptes Rendus Chimie. 2003;6(2) 259-264.

[52] Tronto J, Sanchez KC, Crepaldi EL, Naal Z, Klein SI, Valim JB. Synthesis, Characterization and Electrochemical Study of Layered Double Hydroxides Intercalated with 2-Thiophenecarboxylate Anions. Journal of Physics and Chemistry of Solids. 2004;65(2-3) 493-498.

[53] Moujahid EM, Dubois M, Besse JP, Leroux F. In situ Polymerization of Aniline Sulfonic Acid Derivatives into LDH Interlamellar Space Probed by ESR and Electrochemical Studies. Chemistry of Materials. 2005;17(2) 373-382.

[54] Tronto J, Leroux F, Crepaldi EL, Naal Z, Klein SI, Valim JB. New Layered Double Hydroxides Intercalated with Substituted Pyrroles. 1. In Situ Polymerization of 4-(1H-pyrrol-1-yl)benzoate. Journal of Physics and Chemistry of Solids. 2006;67(5-6) 968-972.

[55] Tronto J, Leroux F, Dubois M, Taviot-Gueho C, Naal Z, Klein SI, Valim, JB. New Layered Double Hydroxides Intercalated with Substituted Pyrroles. 2. 3-(Pyrrol-1-yl)-Propanoate and 7-(pyrrol-1-yl)-Heptanoate LDHs. Journal of Physics and Chemistry of Solids. 2006;67(5-6) 973-977.

[56] Tronto J, Leroux F, Dubois M, Borin, JF, Graeff, CFD, Valim JB, Hyperfine Interaction in Zn-Al Layered Double Hydroxides Intercalated with Conducting Polymers. Journal of Physics and Chemistry of Solids. 2008;69(5-6) 1079-1083.

[57] Tian X, Wei M, Evans DG, Rao G, Yang H. Controlled Polymerization of Metanilic Anion within the Interlayer of NiA1 Layered Double Hydroxide. Clays and Clay Minerals. 2006;54(4) 418-425.

[58] Tian X, Wei M, Evans DG, Rao G, Duan X. Tentative mechanisms for In Situ Polymerization of Metanilic Acid Intercalated in MgAl Layered Double Hydroxide Under Nitrogen Atmosphere. Advanced Materials Research.2006;11-12295-298.

[59] Wei M, Tian X, He J, et al. Study of the In Situ Postintercalative Polymerization of Metanilic Anions Intercalated in NiAl-Layered Double Hydroxides under a Nitrogen Atmosphere. European Journal of Inorganic Chemistry. 2006;17 3442-3450.

[60] Bordonal AC. Materiais Híbridos Orgânico-Inorgânicos: Polímeros Condutores Intercalados em Compostos Lamelares. MS thesis. Universidade de São Paulo; 2012.

[61] Tronto J. Síntese, Caracterização e Estudo das Propriedades de Hidróxidos Duplos Lamelares Intercalados com Polímeros Condutores. PhD thesis. Universidade de São Paulo; 2006.

Chapter 2

Atomic Layer Deposition on Self-Assembled-Monolayers

Hagay Moshe and Yitzhak Mastai

Additional information is available at the end of the chapter

1. Introduction

Atomic layer deposition (ALD) is an advanced technique for growing thin film structures. ALD was developed by Tuomo Suntola and co workers in 1974. At first, the method was called Atomic layer epitaxy (ALE). However, today the name "ALD" is more common. The motivation behind developing ALD was the desire to achieve a technique for creating thin film electroluminescent (TFEL) flat panel displays. [1]- [7]

Several types of materials including metals [8], metal oxides [9], metal nitrides [7] and metal sulfides [10] can be deposited into ALD thin films, depending on the precursors used. ALD advantages are: precise and easy thickness control, superior conformality, the ability to produce sharp interfaces, the substrate size is limited by the batch size and straightforward scale up and repetition of the process.[2-4,6] ALD is appropriate for deposition processes which require angstrom or monolayer level control over coating thickness and/or are maintained on complex topographies of the substrate. No other method for thin film creation can get close to the conformality obtained by ALD.[4] ALD also has several limitations. The Achilles' heel of the method is that ALD is a slow process and therefore is not economic for many industrial processes.[2]

Atomic layer deposition controlled film growth is a significant technology for surface chemistry. In the last four decades, ALD has developed into a system used for depositing thin films in a variety of products. For example, ALD is used in microelectronic production, construction of optical and magnetic devices, flat panel displays, catalysts, and energy conversion including solar cells, utilizing fuel cells, storage batteries or supercapacitors, nanostructures as AFM tips, biomedical purpose and more. [11]

Self Assembled Monolayers (SAMs) are ordered molecular (organic molecules in most cases) assemblies formed by adsorption of molecules on a solid substrate. The surface properties of the surfaces formed are determined by the nature of the adsorbed molecules. [12] A typical surfactant molecule for SAMs is built from three main parts. The first part has a high affinity to the solid surface and is called the "headgroup". The headgroup forms a chemical interaction with the substrate. While adsorbing, the molecules make an effort to adsorb at all surface sites, resulting in a close-packed monolayer. The second molecular part is the alkyl chain. The Van der Waals interactions between these chains cause the SAMs to be ordered. The third part which is exposed at the surface is called the "terminal group". The chain can be terminated with several different groups e.g. CH_3, OH, COOH or NH_2, allowing the SAMs to be applied for the modification of surface properties. Thus, SAMs can modify the surface free energies of the substrates, ranging from reactive, high energies, to passive, low energies. [12],[13]

This book chapter will focus on a new application of ALD as a novel method for thin film deposition on SAMs. Since ALD is very sensitive to surface conditions, it is an ideal method for film deposition on SAMs. Examples for the application of ALD on SAMs are, for instance, surface patterning and selective deposition of thin films. ALD is a very suitable method for the deposition of thin films with three dimensional structures.[8] This book chapter will cover the most recent and novel applications of ALD used for the preparation of chiral nanosized metal oxide films using chiral SAMs.

1.1. Principle of technique

ALD is a Chemical vapor deposition (CVD) process with self-limiting growth and is controlled by the distribution of a chemical reaction into two separate half reactions; the film is done in a growth cycle. Throughout the process, the precursor materials have to be separate. A growth cycle includes four stages: 1) Exposure of the first precursor, 2) purge of the reaction chamber, 3) exposure of the second precursor, and 4) a further purge of the reaction chamber [2,5] (Figure 1). In the first stage, the first precursor reacts with all the sites on the substrate receiving a single molecular layer of the first precursor. The second stage consists of Argon flowing and pumping of the residue of the first precursor to avoid unwanted gas phase reactions between precursors, a reaction which will prevent acceptance of a single molecular layer. In the third stage, the second precursor reacts with one molecular layer of the first precursor to get a single molecular layer of the target material. The fourth stage consists of pumping the residuals of the second precursor [2,5,6]. The cycle ends after four stages. The film thickness is determined by the number of cycles because one cycle deposits one molecular layer (Figure 2). [9] Every stage in the process has to be fully completed before the next stage starts. This means that all the sites on the substrate must react with the precursor and the extra precursor molecules must be removed. The molecular size of the precursor determines the film thickness per cycle. The film density obtained depends on the molecular volume of the precursor- that is to say, a molecule with steric hindrance will probably prevent the formation of a monolayer while small molecules without steric hindrance will allow the formation of a full monolayer. The density of the reactive sites on the substrate is also significant for the nature of the film obtained. One cycle can take from half a second to a few seconds depending on the reactivity between the gas precursors and the solid substrate. In ALD, spontaneous reactions are desired. [2], [5], [6]

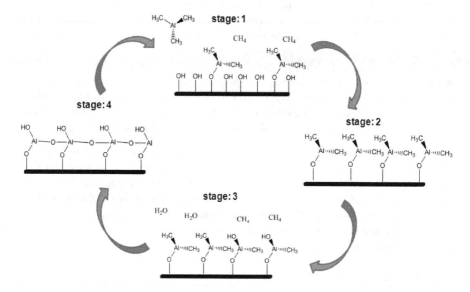

Figure 1. ALD growth cycle includes four stages: 1) Exposure of the first precursor, 2) purge of the reaction chamber, 3) exposure of the second precursor, and 4) a further purge of the reaction chamber.

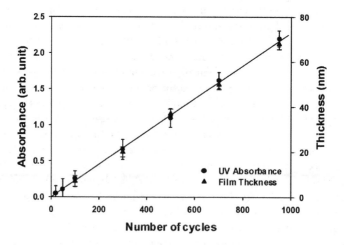

Figure 2. UV absorbance and thickness of the TiO_2 thin films versus the number of ALD cycles. [9]

Depending on the precursors used, ALD can deposit several types of materials, including metals as Pt^{14}, Ru^{15} and Ir^{16}; metal oxides as ZnO^{17}, TiO_2^9, ZrO_2^{18} and HfO_2^{19}; metal nitrides as $Hf_3N_4^{20}$ and $Zr_3N_4^{20}$; metal sulfides as PbS^{10} and Polymers as Polyimide [16]. Using the right precursor is one of the keys for a successful ALD process. Good ALD precursors have to include a number of properties during deposition conditions. First, they have to be stable, evaporable and react with the substrate to completeness. In addition, they must be safe, non toxic and inexpensive. Finally, there should be no etching of the substrate or the growing film and inert volatile byproducts. [2,5-7]

The material type of ALD thin films depends on the precursor type. Normally halides, alkyl compounds, and alkoxides are used as metal precursors. Nonmetal precursors include water, hydrogen peroxide, and ozone for oxygen; hydrides for chalcogens; ammonia, hydrazine, and amines for nitrogen; hydrides for the fifth group in the periodic table. [2], [5]

ALD precursors can be in any state of matter- gas state, liquid state, or solid state. In order to have an effective feeding of precursor molecules to the system, vapor pressure should be high enough. The precursor is also heated sometimes. There must be enough precursors to cover all sites on the substrate surface. [2], [5]- [7]

1.2. Advantages and disadvantages

The ALD technique has a number of advantages: ALD has angstrom or monolayer level control on thickness, the film thickness depends only on the number of reaction cycles. ALD has large area deposition ability, the area size depends only on the ALD chamber size. ALD is a very suitable method for the deposition of thin films with three dimensional structures. [8] As a result, ALD has excellent conformality to substrate surfaces. ALD is a reproducible process, can work on low temperatures and uses highly reactive precursors. The ALD method allows processing of different materials in a continuous process. [2], [5], [7]

ALD's weak point is its slow growth rate; one monolayer is deposited per cycle. The monolayer thickness is a few angstroms; if a cycle takes a few seconds, micron thickness deposition will take a few hours. Consequently, ALD is not a useful method for many applications. The growth of films in micrometer size takes too long for ALD to be an economic industrial process. This problem is sometimes overcome by using a big chamber to be able to contain many substrates per batch, but a single wafer process is still more ideal. ALD is an unselective process. Generally, precursor molecules react with all surfaces. In order to achieve selectivity control, pretreatment is necessary. The places which should not be deposited on have to be passivated. As a chemical process, ALD has a risk of impurities. The impurities can come from gas precursors and/or a carrier gas, the process requires material with a high degree of cleanliness. Impure chemicals can lead to the incorporation of impurities and to the growth of poor quality films. [2], [5], [7]

1.3. ALD process at low temperature

The ability to perform ALD at low temperatures (ALD-LT) is very important. It is critical for ALD on SAMs and it is the subject of this chapter. SAMs as well as polymers or biological samples are thermally sensitive materials. At high temperatures, they decompose. [3] In the case of SAMs, there is also disabsorption from the surface. Inter-diffusions of materials occur at high temperature processes, it has a devastating effect on nano-structured devices. ALD at low temperatures avoids these effects. To carry out ALD-LT, a catalyst is sometimes used [3], [21], [22] although there are reactions that occur without catalysts. [3], [23] Nanostructures of biological structures have very interesting effects. For example, a lotus leaf shows highly hydrophobic behavior due to its nanostructures. The coat of the lotus leaf can be copied by ALD-LT, achieving similar effects. ALD-LT was also used on a tobacco mosaic virus (TMV) on protein spheres [24] and on cellulose fibers from filter paper.[3], [25]

2. ALD on self-assembled-monolayer

The use of ALD for depositing thin films onto different SAMs has great potential applications. SAMs are thin organic films which form spontaneously on solid surfaces. The SAM head group has to connect to the substrate strongly enough for stable monolayers to form. Typical SAM head groups are alkanethiols [X-(CH_2)n–SH] which are formed on metal surfaces such as Ag, Au, and Cu, and alkyltrichlorosilanes [X-$(CH_2)_n$–$SiCl_3$] formed on SiO_2, Al_2O_3, and other oxide surfaces. [26]- [28]

In general, SAMs are formed by immersing the substrates into a solution comprising the precursor molecules or by bringing the SAM precursors to the substrate surface as vapors. [16] SAMs are well known to modify the physical and chemical properties of surfaces. The surface features can be controlled by using the appropriate SAM. Potential applications include control of wetting and friction behaviors, passivating layers, protection of metals against corrosion, preparation of chiral surfaces, molecular electronics, chemical sensing, soft lithography and more. [10], [26], [29] Figure 3 shows water droplet angle measurements demonstrating the formation of hydrophobicity by ODTS (octadecyltrichlorosilane) SAM on originally hydrophilic SiO_2. [26]

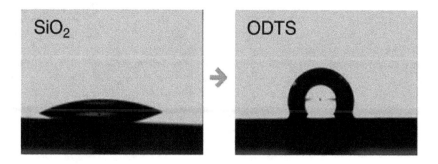

Figure 3. Contact angle measurements showing the control of surface energy by ODTS SAM before and after treatments on SiO_2 substrate. [26]

ALD onto SAMs is interesting because the ordered structure of the monolayer can act as a template for the growth of structured thin films. The SAM can be used to engineer the properties of the interface to the original substrate, when the ALD coatings are protecting the SAM. [30]

The thermal stability of the SAMs under the ALD process conditions is very important because of the potential loss of ordering at elevated temperatures. In order to maintain the SAM order during the growth of the ALD film, the ALD process must be done under conditions which are compatible with the thermal budget of the underlying SAM film. [30]

2.1. Area-Selective ALD on SAM

Patterned SAMs are commonly used as growth-preventing masks for selective-area ALD. Selective-area ALD is the growth of thin films on the substrate surface on designated sites only. Selective-area ALD requires that the chosen regions of the surface are inert to ALD precursors. In this case the function of the SAM is to protect the surface against deposition. ALD grows only on areas without a SAM, on the desired sites of the surface [16]. High-resolution patterns can be created by printing SAMs by means of soft lithography, [31]- [33] or by removing SAMs using electron beams, [34] ion beams, photolithography, [35] or scanning probe microscopy. [36] Generating patterned SAMs in the most economical way is a critical necessity for using patterned SAMs in advanced applications. Photolithography can transfer an entire pattern on a photomask to a SAM at a given time. Therefore, it is the most practical among various patterning methods. [18]

During the past decade, several groups have used SAMs as a chemical resist to block various ALD precursors, including ZnO [17] [37], TiO$_2$[9], [38]- [40], ZrO$_2$ [18], HfO$_2$ [19], [41]- [45], Ru [15], Ir [16], [40], [46], Pt [14], [42], [45], [47]- [49], PbS [10] and Polyimide [16]. In 2001 Yan et al. reported on selective area ALD growth of ZnO, they used a microcontact printing (or µCP) with poly(dimethylsiloxane) (PDMS) stamp as a soft lithographic technique. PDMS creates a hydrophobic surface in the stamped area, leaving the ink-free hydrophilic surface unmodified. The pattern consists of arrays of cylinders having cross sectional diameters ranging from 1.0– 40 µm with center-center distances of 100 µm. As ALD precursors they used diethylzinc (DEZ) and deionized water. The deposition process for ZnO consists of two self-limiting chemical reactions, repeated in alternation (ABAB...). Each AB reaction cycle deposits a single monolayer of ZnO, [50] as shown in Eqs. (1) and (2), where asterisks indicate the outermost surface functional groups.

$$A: Zn-OH^* + Zn(CH_2CH_3)_2 \rightarrow Zn-O-Zn-CH_2CH_3^* + C_2H_6 \uparrow \qquad (1)$$

$$B: Zn-CH_2CH_3 + H_2O \rightarrow Zn-OH^* + C_2H_6 \uparrow \qquad (2)$$

The deposition was carried out at a substrate temperature of 125 °C. The exposure times for DEZ and water vapor were 0.7 sec and 0.5 sec, respectively. ZnO nucleation and growth do not occur on the; 2 nm thick SAM-patterned areas, but only on the bare, hydrophilic unpatterned areas as illustrated in Figure 4. [17]

Park et al. reported on a patterning method of TiO$_2$ thin films using microcontact printing of alkylsiloxane SAMs, followed by selective atomic layer deposition of the TiO$_2$. Park et al. approach consists of two key steps. First, the patterned alkylsiloxane SAMs were formed by using microcontact printing. Second, the TiO$_2$ thin films were selectively deposited onto the SAM-patterned Si substrate by atomic layer deposition. [9]

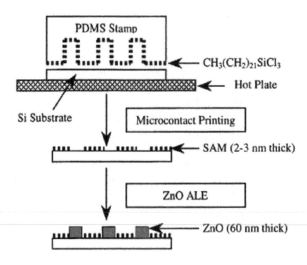

Figure 4. Schematic outline of the patterning and deposition procedures used for selective area ZnO thin film growth. [17]

Seo et al. reported on a patterning method for TiO$_2$ thin films using microcontact printing of alkanethiolate SAMs on gold, followed by selective atomic layer deposition of the TiO$_2$. Seo et al. approach consists of three key steps. First, patterned CH$_3$-terminated alkanethiolate SAMs on gold were formed by using microcontact printing. Second, the remaining regions of gold were coated with OH-terminated alkanethiolate SAMs. Third, the TiO$_2$ thin films were selectively deposited onto the SAM-patterned gold substrate by atomic layer deposition. [38] Both groups used PDMS stamp as a soft lithographic technique and same conditions for the preparation of TiO$_2$ thin films. As ALD precursors they used Titanium isopropoxide (Ti(OPri)$_4$) and deionized water. The Ti(OPri)$_4$ and water were evaporated at 80 and 20 °C, respectively. The cycle consisted of 2 sec exposure to Ti(OPri)$_4$, 5 sec Ar purge, 2 sec exposure to water, and 5 sec Ar purge. The total flow rate of the Ar was 20 sccm. The TiO$_2$ thin films were grown at 100 °C under 2 Torr. The deposition process for TiO$_2$ also consists of two self-limiting chemical reactions, repeated in alternation (ABAB...). Each AB reaction cycle deposits a single monolayer of TiO$_2$ [9], [38]. Figure 5 illustrates AFM images and cross sections of micropatterned TiO$_2$ thin films, which were selectively deposited onto the monolayer-patterned gold substrate by ALD. The patterned SAMs showed high selectivity for TiO$_2$ ALD;

hence, the patterns of the TiO_2 thin films were defined and directed by the patterned SAMs generated with microcontact printing. The TiO_2 thin films are selectively deposited only on the regions exposing the OH groups of the MUO-coated gold substrates, because the regions covered with the ODT monolayers do not have any functional group to react with ALD precursors. These AFM images clearly show that the patterned TiO_2 thin films retain the dimensions of the patterned SAMs used as templates with no noticeable line spreading. [38]

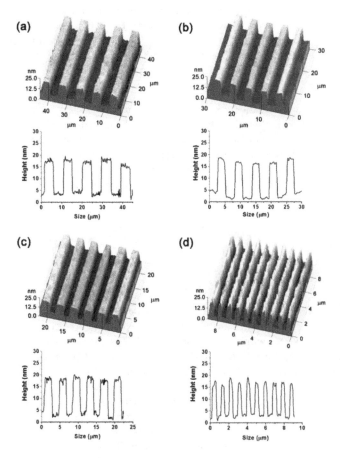

Figure 5. AFM images and cross sections of the patterned TiO_2 thin films generated by using selective ALD on the SAM patterned gold substrates: (a) 3.7 μm lines with 5.6 μm spaces, (b) 1.9 μm lines with 3.7 μm spaces, (c) 1.8 μm lines with 1.9 μm spaces, (d) 0.5 μm lines with 0.4 μm spaces [38].

The successful use of poly(methyl methacrylate) (PMMA) [39] and octadecyltrichlorosilane (OTS) SAMs [40] as a mask layer to obtain the direct patterned deposition of TiO_2 films has been reported.

Lee and Sung reported on a fabrication method using photocatalytic lithography of octade-cylsiloxane SAMs, followed by selective deposition of ZrO_2 thin films using ALD. Lee et al. approach consists of three key steps. First, the alkylsiloxane SAMs were formed by immersing Si substrate in alkyltrichlorosilane solution. Second, photocatalytic lithography using a quartz plate coated with patterned TiO_2 thin films was done to prepare patterned SAMs of alkylsi-loxane on the Si substrate. The patterned SAMs of the octade octadecylsiloxane on the Si substrate were made by using the quartz plate coated with the patterned TiO_2 thin films under UV irradiation in air. The photocatalytic lithography is based on the fact that the decomposition rate of the alkylsiloxane monolayers in contact with the TiO_2 is much faster than that with the SiO_2 under UV irradiation in air. These patterned SAMs define and direct the selective deposition of the ZrO_2 thin films. Third, ZrO_2 thin films were selectively deposited onto the SAMs-patterned Si substrate by ALD. A ZrO_2 thin film was selectively deposited using $Zr(OC(CH_3)_3)_4$ and water as ALD precursors. [18]

Chen et al. investigated a series of self assembled molecules as monolayer resists for HfO_2 atomic layer deposition. A series of n-alkyltrichlorosilanes of chain lengths ranging from 1 to 18 carbon atoms was used to form self-assembled monolayers on the oxide-covered silicon substrates. The ALD precursors for HfO_2 deposition were hafnium tetrachloride ($HfCl_4$) and water. The HfO_2 ALD process includes two self-limiting chemical reactions, repeated in alternating ABAB sequences shown in Eqs. (3) and (4), where asterisks indicate the outermost surface functional groups.

$$A : Hf - OH^* + HfCl_4 \rightarrow Hf - O - HfCl_3^* + HCl \uparrow \qquad (3)$$

$$B : Hf - Cl^* + H_2O \rightarrow Hf - OH^* + HCl \uparrow \qquad (4)$$

Each AB reaction cycle produces an HfO_2 layer terminated by hydroxyl groups, with the hydrochloride byproduct pumped away. After each exposure, the reaction chamber and the gas manifold were purged with nitrogen to avoid possible gas-phase reactions and to eliminate the possible physisorption of the precursors on the substrates. Deposition was carried out at a substrate temperature of 300 °C. The exposure times for $HfCl_4$ and water vapor were both 2 sec, followed by 3 min of nitrogen purging after each precursor was introduced into the chamber. Chen et al. explained the blocking mechanism by three important factors that influence the blocking efficiency of the monolayer organic films: chain length, tailgroup structure, and headgroup reactivity. This investigation shows that to achieve satisfactory deactivation toward the ALD process, it is crucial to form densely packed, highly hydrophobic organic monolayers. This in turn requires deactivating agents with high reactivity, low steric effect tail groups, and minimum chain length. [19]

Park et al. presented a method that combines SAM passivation and high-k dielectric deposi-tion. Tetradecyl-SAM is formed on a Ge (100) surface via a thermal method. Part of the SAM is then removed by proper annealing, and a HfO_2 film is deposited by ALD. The system development was based on a previous research on the electrical properties of SAMs on Ge surfaces showing that SAM/Ge interfaces are electrically stable compared to Ge surfaces covered with native oxide. [51] Therefore, a combination of SAM passivation on a Ge surface with high-k gate dielectric deposition is suggested for scaling down gate oxides. [43]

Liu et al. created sub- 10 nm patterns of high dielectric constant (high-k) HfO_2 on Si substrate, by combining the use of the reassembled S-layer proteins as nanotemplates and an area-selective ALD process. To realize area-selective ALD of metal oxide-based high-k material nanopatterns into the nanotemplates composed of protein architectures, it is necessary to modify the S-layer proteins to introduce different surface functional groups upon them and the silicon substrate surface used in Liu et al. study. As a result, ALD only happens on the Si substrate and cannot take place on the modified surface of the S-layer proteins. ODTS used as an effective monolayer resist on a hydrophilic SiO_2 surface toward ALD of HfO_2, was chosen to modify the surface of the S-layer proteins but not the Si surface. Specifically, the ODTS-modified S-layer proteins are terminated with aliphatic chains (R=$(CH_2)_{17}CH_3$), while the Si surface exposed through the pores defined by the protein units is terminated with -OH or -H functional groups. Since atomic layer deposition has been achieved ideally on surfaces with -OH groups and with an incubation time on surfaces with -H groups, it is therefore feasible to achieve area-selective ALD on a surface with a contrast between aliphatic groups and -OH/-H terminations. [52], [53] ODTS-modified S-layer protein nanotemplate was selectively removed by thermal annealing. Therefore, S-layer proteins reassembled on Si substrate acted as a promising nanotemplate for the sub-10-nm nanopatterning of high-k oxides for future Metal Oxide Semiconductor Field Effect Transistor (MOSFET) applications. [44]

Park et al. demonstrated selective deposition of Ruthenium using contact printed self-assembled monolayer resists by selective area atomic layer deposition. Ruthenium is of interest for advanced metal/oxide/semiconductor (MOS) transistor gate electrodes to reduce polysilicon depletion effects and as nucleation layer for copper interconnect layers. [54] Ruthenium is considered as a viable candidate for p-type MOS devices because it has a vacuum work function near the conduction band edge of silicon, good thermal stability, and low resistivity of the oxidation phase. [55] Selective deposition enables direct formation of $Ru/HfO_2(SiO_2)/Si$ capacitor stacks, and the effective work function of ALD Ru is characterized on HfO_2 and SiO_2 dielectrics. They used PDMS stamps and OTS SAMs to prepare the patterned organic monolayer. ALD Ru was carried out using bis-(cyclopentadienyl)rutheniumg ($RuCp_2$) as a precursor and dry oxygen. $RuCp_2$ is solid at room temperature with vapor pressure of ~10 mTorr at the bubbler temperature of 80 °C. The ALD chamber was evacuated to $5*10^{-6}$ Torr, and the precursor and oxidant gases were introduced into the reactor in separate pulses (3 and 6 sec, respectively) with a 20 sec Ar purge between each reactant. Argon was also used as a carrier gas for the $RuCp_2$ pulse. [15]

Färm et al. reported on selective deposition of Iridium by using octadecyltrimethoxysilane (ODS), SAMs prepared from gas phase using a process where water-vapor pulses were given alternately with ODS. SAMs were patterned by a simple lift-off process. [46] In another work, narrow lines of OTS was printed by PDMS stamp which had 1.5 μm wide print lines and 1.5 μm wide spaces between. They also presented the passivation of copper surfaces using 1-dodecanethiol ($CH_3(CH_2)_{11}SH$) SAMs against iridium ALD growth. 1-dodecanethiol was chosen as a SAM precursor because it has relatively long carbon chain, it is liquid and volatile enough so that SAM were prepared from the vapor phase using moderate heating. [16] Iridium was grown only on non-SAM areas at 225 °C from Ir (pentanedione)$_3$ and O_2. [16], [40], [46]

Chen and Bent reported on deposition of Pt for the positive patterning area-selective ALD. Pt is a promising electrode material for dynamic random-access memories because of its high chemical stability in an oxidizing atmosphere and its excellent electrical properties. [56], [57] It is also a promising gate metal candidate owing to its high work function (5.6 eV) and compatibility with high-k dielectrics. [58] In the Chen and Bent paper, a deposition of Pt occurs on a SiO_2 film, providing a model process for the deposition of a gate metal on a dielectric. They used 1-octadecene as a monolayer, which undergoes a hydrosilylation reaction selectively on the hydride surface. Following monolayer attachment onto oxide patterned silicon wafers, Pt thin films were selectively deposited onto the substrates. ALD of a Pt thin film was carried out using methylcyclopentadienyl(platinum)trimethyl ($CH_3C_5H_4Pt(CH_3)_3$) and dry air. Exposure times for the Pt precursor and air were 3 and 2 sec, respectively, followed by a 60 and 45 sec N_2 purge after each precursor was introduced into the chamber. The Pt ALD process includes two self-limiting chemical reactions, repeated in the alternating ABAB sequences shown in Eqs. (5) and (6), where asterisks indicate the outermost surface functional groups and OBP, H_2O, CO_2 are reaction byproducts. [14], [42]

$$A: Pt + O_2 \rightarrow Pt - O^* \tag{5}$$

$$B: CH_3C_5H_4Pt(CH_3)_3 + Pt - O^* \rightarrow Pt + CO_2 \uparrow + H_2O \uparrow + OBP \uparrow \tag{6}$$

Figure 6 illustrates the Auger electron spectroscopy (AES) analysis of the patterned lines at higher spatial resolution. Figure 6a shows a SEM image of the patterned lines used for the study. In the SEM image, the oxide and the deactivated hydride regions (areas 1 and 2 in Fig. 6.a, respectively) were chosen for AES compositional analysis. The Auger survey scans shown in Figure 6.b reveal that, in the deactivated lines (area 2), the Pt signal is below the AES detection limit (0.5 %), whereas significant Pt is seen in area 1. AES line-scan images that compare the amounts of C and Pt as a function of position are displayed in Figure 6.c. A cross-sectional line (similar to the dashed line shown in Fig. 3a) was obtained perpendicular to the patterned lines. The C and Pt spectra clearly show the alternation as expected, and the edges of the Pt lines are sharp. [14]

Jiang and Bent reported on area selective atomic layer deposition of Platinum on Yttria stabilized zirconia (YSZ) substrates using microcontact printed SAMs. Jianga and Bentb technique can be used to deposit Pt on an YSZ solid oxide electrolyte for the catalyst in solid oxide fuel cell (SOFC). Pt is the catalyst used for a number of reactions, including the O_2 reduction reaction at the cathode of a SOFC, and is especially useful at the lower operating temperatures below (600 °C) that are desired for integratable fuel cell systems. [48], [49]

Lee et al. reported on capability of SAMs to block the deposition of PbS thin films by ALD. ODTS SAMs were chosen to modify the surface termination because of their ability to

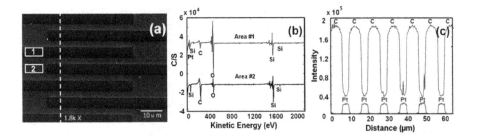

Figure 6. AES analysis on a patterned structure after the area-selective Pt ALD process: a) SEM image of the patterned area, b) AES selected-area survey composition scan, and c) AES defined line scan. [14]

deactivate ALD reactions as well as their good chemical and thermal stability. Microscale patterns of ALD PbS with high spatial and chemical selectivity were fabricated on ODTS patterned Si/SiO₂ substrates. ODTS was selectively grown only on the oxide patterns defined by the photolithography and deactivated PbS deposition during the ALD process. Hence, materials were selectively deposited by the ALD process only where ODTS was not present. The ALD precursors used were bis(2,2,6,6-tetramethyl-3,5-heptanedionato)lead(II) (Pb(tmhd)2) and H₂S. The base pressure of the ALD chamber was 50 mTorr. The substrate temperature was maintained at 160 °C, and the precursor was sublimated at 140 °C. [59] The PbS ALD process includes two self-limiting chemical reactions, repeated in an alternating ABAB sequence. Lee et al. postulated the following ligand-exchange reactions, which are typical of ALD half-reaction chemistry. In both reactions, a gas-phase precursor molecule reacts with the surface functional species and saturates the entire surface in a self-limiting manner. Each AB reaction cycle produces a PbS layer terminated by sulfhydryl groups, with the corresponding byproducts pumped away. [10]

Färm et al. reported on passivation of copper surfaces for selective-area ALD using 1-dodec-anethiol SAMs against polyimide ALD growth. Polyimide is a new material for selective-area ALD and has potential applications as an insulating material in copper interconnects. As test substrates, silicon with evaporated copper dots was used. SAMs were prepared on the copper surfaces from the vapor phase. Polyimide was deposited from 1,2,3,5-benzenetetracarboxylic anhydride (pyromellitic dianhydride) and 4,4-oxydianiline at 160 °C. [16]

SAMs have typically been created by dipping the solid substrates into a solution containing the precursor molecules. The vapor process in preferred and involves preparing SAMs by bringing the precursors to the substrate surface as vapors. The vapor process has some advantages over the liquid process, e.g., when SAMs have to be formed on three-dimensional structures. The vapor process can prevent problems related to the absorption of liquids into the porous structures. [16] The vapor-phase process also requires fewer precursors than the liquid-phase processes. Moreover, the aggregation of the precursor molecules prior to deposition on the substrate surface, which can cause defects in the arrangement of the SAMs in the liquid-phase process, is significantly reduced using the vapor-phase process. Aggregated precursors had lower vapor pressures than the single-molecule precursors and thus were rarely vaporized. [45] The vapor-phase SAM formation can be carried out in a vacuum system allowing easier combination with the ALD reactor. In principle, SAM formation can be performed in the ALD reactor itself. [45] When SAMs are prepared as an initial stage of the ALD process, the patterning of the SAMs has to be done by relying on the chemical selectivity of the SAM formation. [16] Silane [40], [45], [46] and thiol [16] SAMs has been formed from the vapor phase for selective-area ALD of TiO_2 [36], HfO_2 [45], Ir [16], [40], [46], Pt [45] and Polyimide. [16]

The common way to block ALD by using SAMs is limited when the height of the deposited inorganic film exceeds the height of the self-assembled monolayer (~2 nm). In that case the growth will not be area-selective anymore near the interface where the already deposited inorganic film meets the end of the alkyl tails. Near that interface, the ALD reactants are able to adsorb on the inorganic film.

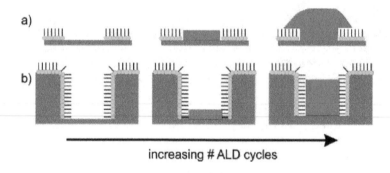

increasing # ALD cycles

Figure 7. (a) Conventional area-selective ALD in which the substrate is planar and contains patterns of self-assembled monolayers. With increasing number of deposition cycles there occurs also sideways film growth originating from adsorption of ALD reactants on the previously deposited ALD film. b) Blocking the lateral ALD growth independent of deposited film thickness by combining surface modification and topographical features. [37]

The inorganic film is not confined anymore to the original pattern of the SAM and the lateral dimension of the film will increase when more ALD cycles are carried out (see Figure 7.a). Robin at el. have shown a new concept to enable construction of nanoscale lateral structures by area-selective ALD. The concept is based on providing chemical inertness by surface modification combined by nanoscale topographical structures (Figure 7.b). Whereas surface modification, as traditionally used in area-selective ALD, is only a chemical barrier for film growth, Robin et al. shows that the topographical structures are also a physical barrier for film growth. Their concept allows ALD synthesis of constructs that have lateral dimensions many times smaller than the film thickness. Robin et al. used cicada wings as a prototypical example from nature; however, their concept can be also applicable on other types of designed substrates that combine surface modification (including SAMs) with nanoscale topographies. [37]

2.2. Surface study by ALD on SAM

Lee et al. studied a surface free energy by atomic layer deposition of TiO_2 on mixed SAMs. They studied ALD growth modes as a function of surface free energy. [60] Mixed SAMs have been used to modify the surface free energy of the Si substrates. By using solutions containing two different silanes, it formed SAMs containing mixtures of them. The influence of the surface free energy of the Si substrates on the growth modes of TiO_2 thin films has been studied with AFM, XPS and contact-angle analysis. Mixed SAMs with several surface compositions of H_3C-Si and HO-Si were formed on the Si substrates. The surface free energy of the SAM-contact samples was derived from the contact-angle data by using water and diiodomethane, as shown in Table 1. [60]

HO–Si : CH$_3$–Si	contact angle θ [°]		surface free energy [mN/m]		
	water	diiodomethane	γ_s	γ_s^d	γ_s^p
1 : 0	40	25	64	41	22
4 : 1	51	34	56	38	18
2 : 1	62	40	49	36	13
1 : 1	70	45	43	34	9
1 : 2	78	50	38	32	6
1 : 3	89	55	33	31	3
0 : 1	108	60	29	29	0

Table 1. Contact angle (θ) and surface free energy (γ_s) of Si substrates coated with SAMs (d: dispersive part, p: polar part). [60]

The surface free energy of the mixed SAM-contact samples, ranging from 64 to 29 mN/m, appears to be determined primarily by the surface composition of the of H_3C-Si and HO-Si, which means that the surface free energy of solid substrate can be controlled by mixed SAMs. The TiO_2 thin films were grown on the mixed SAM-coated Si substrates by atomic layer deposition from titanium isopropoxide and water. The ALD growth mode of the TiO_2 film changes as function of the surface free energy of the Si substrates, and the surface free energy can be modified by changing the ratio of the components of the mixed SAMs. A two-dimensional growth mode is observed on the SAM-coated substrates with high surface free energies. As the surface free energy decreases, a three-dimensional growth mode begins to dominate. From the results, Lee et al. have found that the mixed SAMs can be used to control the growth modes of the atomic layer deposition by modifying the surface free energy of the substrates. [60]

Xu and Musgrave used density functional theory (DFT) for investigated surface reactions between trimethylaluminum (TMA) as precursor for alumina and SAMs terminated with different functional groups. [30] They show that the reaction of TMA and the -OH-terminated SAM is favored both thermodynamically and kinetically over the reaction with -NH_2- and -CH_3-terminated SAMs. Reactions on the -NH_2-terminated SAM form more stable complex intermediates; however, because the ligand exchange barrier is large, the precursors are trapped in the adsorbed complex state. Furthermore, although there is a thermodynamic driving force for this reaction, the reaction is relatively slow compared to the -OH-terminated case and desorption of the precursor is favored over ligand exchange. In the case of the -CH$_3$-terminated SAM, there is no thermodynamic driving force for the reaction and the reaction barrier is large. The reaction path and predicted energetics for reactions of TMA and -OH/NH$_2$/CH$_3$-terminated SAMs as shown in Figure 8. [30]

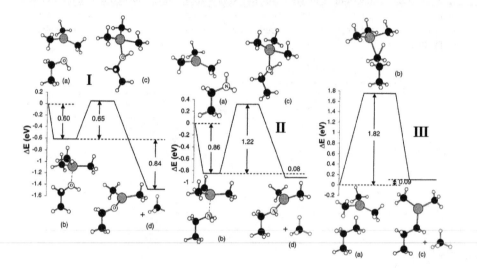

Figure 8. I) Reaction path and predicted energetics for reactions of TMA and -OH-terminated SAM. The stationary points correspond to (a) CH$_3$CH$_2$OH + TMA, (b) complex TMA•OHCH$_2$CH$_3$, (c) transition state, and (d) CH$_3$CH$_2$O-Al(CH$_3$)$_2$+ CH$_4$. II) Reaction path and predicted energetics for reactions of TMA and -NH$_2$-terminated SAM. The stationary points correspond to (a) CH$_3$CH$_2$NH$_2$ + TMA, (b) complex TMA•NH$_2$-CH$_2$CH$_3$, (c) transition state, and (d) CH$_3$CH$_2$NH-Al(CH$_3$)$_2$ +CH$_4$. III) Reaction path and predicted energetics for reactions of TMA and -CH3-terminated SAM. The stationary points correspond to (a) CH$_3$CH$_2$CH$_3$ + TMA, (b) transition state, and (c) CH$_3$CH$_2$CH$_2$-Al(CH$_3$)$_2$ + CH$_4$. [30]

The energetics of the reactions does not depend on the length of the SAM using ethyl and pentyl groups as models. [30] After the initial TMA adsorption on the -OH-terminated SAM, the second half-reaction of ALD growth of Al$_2$O$_3$ (Al-CH$_3$*→Al-OH*) is calculated and the mechanism and energetics are consistent with their previous results for ALD of Al$_2$O$_3$ using TMA and water. [61] Because these adsorption reactions are highly localized, the conclusions are not only limited to the effect of surface functionalization on ALD reactions on SAMs; they can also be extended to reactions on other substrates and to ALD reactions involving other precursors which form dative-bonded complexes. [30]

Lee et al. used DFT simulation for study reactions between Pb(tmhd)$_2$ precursor to ODTS SAMs and SiO$_2$, the results showed an increased activation barrier and a higher overall reaction energy for the Pb(tmhd)$_2$ precursor on an ODTS-terminated substrate than on a SiO$_2$ surface. [10]

2.3. ALD on chiral SAMs

H. Moshe et al. proposed a new innovative type of stable chiral nanosized metal oxide surfaces. [62] The structure and chirality of this type of chiral surface is based on chiral self-assembled monolayers (SAMs) coated with nanosized films of metal oxide materials deposition by ALD. The idea underlying this new design of nano-chiral surfaces is that the ceramic nanolayers coating the chiral SAMs protects the chiral SAMs that would otherwise be destroyed under the reactions conditions, thereby preserving their enantioselective nature. In Figure 9, the overall structure of the new nanoscale hybrid chiral surfaces based on chiral SAM and ceramic nanolayers is shown.

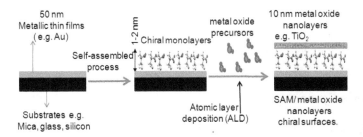

Figure 9. Illustration of the design and the synthesis paths for the new chiral SAM/ceramic nanolayers surfaces.

In their study, H. Moshe et al. used TiO_2 to form the protective nanolayers for the chiral SAM since its synthesis does not demand high temperatures that may harm the chiral SAMs. In the research, they utilize the atomic layer deposition (ALD) technique since it provides excellent thickness control and produces very dense and uniform layers. The first step in the synthesis of this type of nano-chiral surface requires the preparation of chiral SAMs. For the chiral SAMs preparation, they used enantiomers of cysteine and glutathione. TiO_2 films were grown by ALD using $Ti(N(CH_3)_2)_4$ and water as the precursors. Their research focuses on evidence of the chirality of the SAM/metal oxide nanosurfaces. Generally, several techniques [63] can be used to study the chiral nature of nano-sized surfaces such as chiral AFM, STM, second-harmonic generation (SHG) and isothermal titration calorimetry. [64] However, due to the unique structure of their nanosized chiral surfaces, they are rather limited in the techniques that can be used to prove the chirality of these surfaces. In their work, H. Moshe et al. have selected several techniques namely quartz microbalance (QMB), second-harmonic generation circular-dichroism (SHG CD) spectroscopy, enantioselective crystallization and chiral adsorption measurements as the methods to study the chirality of the SAM/TiO_2 nanolayer surfaces.

3. Conclusions

In this book chapter the preparation, properties and applications of ALD as a novel method for thin film deposition on Self Assembled Monolayers have been briefly reviewed. We have reviewed a selective-area atomic layer deposition of a variety of materials such as metals, metal oxides, and polymers. First, we presented a brief introduction reviewing the ALD method and principle of operation. Second, we discussed the ability of SAMs to shape the surface of the substrate before the ALD deposition stage. ALD is very sensitive to surface conditions and therefore offers an ideal method for film deposition. Third, we reviewed procedures for properties and applications of ALD on SAMs. We included a variety of molecules and materials and different conditions used for atomic layer deposition. Fourth, we discussed studies which used ALD on SAMs in order to learn surface properties. Finally, a novel application of ALD for the preparation of chiral nanosized metal oxide films using chiral SAMs was discussed. Future work is aimed at the modification and functionalization of surfaces by SAMs used as templates for ALD.

ALD is a technique with high control capabilities. SAMs are a simple and versatile method used for surface design. Integration of the ALD technique and the SAM method can increase the ability to study and engineer substrate surfaces.

Acknowledgements

H. Moshe would like to acknowledge the Department of Chemistry, Bar Ilan University for funding.

Author details

Hagay Moshe and Yitzhak Mastai

Department of Chemistry and the Institute of Nanotechnology, Bar Ilan University, Israel

References

[1] Suntola, T, & Antson, J. (1977). Method for producing compound thin films. U.S. Patent #4,058,430, Issued Nov. 25

[2] Leskela, M, & Ritala, M. (2003). Atomic layer deposition chemistry: recent developments and future challenges. *Angewandte Chemie International Edition,* , 42

[3] Knez, M, Niesch, K, & Niinisto, L. (2007). Synthesis and surface engineering of complex anostructures by atomic layer deposition. *Advanced Materials,* , 19

[4] George, S. M. (2010). Atomic layer deposition: an overview. *Chemical Reviews,* , 110

[5] Ritala, M, & Leskelä, M. (2002). Atomic layer deposition. *Handbook of Thin Film Materials,* Ed.: Nalwa, H. S.) Academic Press, San Diego, , 1, 103-159.

[6] Niinistö, L, Nieminen, M, Päiväsaari, J, Niinistö, J, Putkonen, M, & Nieminen, M. (2004). Advanced electronic and optoelectronic materials by Atomic Layer Deposition: An overview with special emphasis on recent progress in processing of high-k dielectrics and other oxide materials. *Physica Status Solidi (a)*, , 201

[7] Becker, J. S. (2002). Atomic layer deposition of metal oxide and nitride thin films. Ph.D. dissertation,Harvard University.

[8] Aaltonen, T. (2005). Atomic layer deposition of noble metal thin films. Ph.D. dissertation, University of Helsinki.

[9] Park, M. H, Jang, Y. J, Sung-suh, H. M, & Sung, M. M. (2004). Selective atomic layer deposition of titanium oxide on patterned self-assembled monolayers formed by microcontact printing. *Langmuir*, , 20

[10] Lee, W, Dasgupta, N. P, Trejo, O, Lee, J, Hwang, R, Usui, J, Prinz, T, & Area-selective, F. B. Atomic Layer Deposition of Lead Sulfide: Nanoscale Patterning and DFT Simulations. *Langmuir*, , 26

[11] Tanskanen, J. T, Bakke, J. R, Bent, S. F, & Pakkanen, T. A. (2010). ALD growth characteristics of ZnS films deposited from organozinc and hydrogen sulfide precursors. *Langmuir*, , 26

[12] Zaera, F. (2012). The surface chemistry of atomic layer depositions of solid thin films. *The Journal of Physical Chemistry Letters*, , 3

[13] Ulman, A. (1996). Formation and structure of self-assembled monolayers. *Chemical Reviews*, , 96

[14] Chen, R, & Bent, S. F. (2006). Chemistry for positive pattern transfer using area-selective atomic layer deposition. *Advanced Materials*, , 18

[15] Park, K. J, Doub, J. M, Gougousi, T, & Parsons, G. N. (2005). Microcontact patterning of ruthenium gate electrodes by selective area atomic layer deposition. *Applied Physics Letters*, , 86

[16] Färm, E, Vehkamäki, M, Ritala, M, & Leskelä, M. (2012). Passivation of copper surfaces for selective-area ALD using a thiol self-assembled monolayer. *Semiconductor Science and Technology*, , 27

[17] Yan, M, Koide, Y, Babcock, J. R, Markworth, P. R, Belot, J. A, Marks, T. J, & Chang, R. P. H. (2001). Selective-area atomic layer epitaxy growth of ZnO features on soft lithography-patterned substrates. *Applied Physics Letters*, , 27

[18] Lee, J. P, & Sung, M. M. (2004). A new patterning method using photocatalytic lithography and selective atomic layer deposition. *Journal of the American Chemical Society*, , 126

[19] Chen, R, Kim, H, Mcintyre, P. C, & Bent, S. F. (2005). Investigation of self-assembled monolayer resists for hafnium dioxide atomic layer deposition. *Chemistry of materials*, , 17

[20] Becker, J. S, Kim, E, & Gordon, R. G. (2004). Atomic layer deposition of insulating hafnium and zirconium nitrides. Chemistry of materials, , 16

[21] Gasser, W, Uchida, Y, & Matsumura, M. (1994). Quasi-monolayer deposition of silicon dioxide. Thin Solid Films , 250

[22] Klaus, J. W, Sneh, O, & George, S. M. (1997). Growth of SiO_2 at room temperature with the use of catalyzed sequential half-reactions. Science, , 278

[23] Luo, Y, Slater, D, Han, M, Moryl, J, & Osgood, R. M. (1997). Low-temperature, chemically driven atomic-layer epitaxy: In situ monitored growth of CdS/ZnSe(100). Applied Physics Letters, , 71

[24] Knez, M, Kadri, A, Wege, C, Gösele, U, Jeske, H, & Nielsch, K. (2006). Atomic layer deposition on biological macromolecules: metal oxide coating of tobacco mosaic virus and ferritin. Nano Letters, , 6

[25] Kemell, M, Pore, V, Ritala, M, Leskelä, M, & Linden, M. (2005). Atomic layer deposition in nanometer-level replication of cellulosic substances and preparation of photocatalytic TiO_2/cellulose composites. Journal of the American Chemical Society, , 127

[26] Kim, H, Lee, H. B. R, & Maeng, W. G. (2009). Applications of atomic layer deposition to nanofabrication and emerging nanodevices. Thin Solid Films, , 517

[27] Laibinis, P. E, Whitesides, G. M, Allara, D. L, Tao, Y. T, Parikh, A. N, & Nuzzo, R. G. (1991). Comparison of the structures and wetting properties of self-assembled monolayers of n-alkanethiols on the coinage metal surfaces, copper, silver, and gold. Journal of the American Chemical Society, , 113

[28] Blum, A. S, Kushmerick, J. G, Long, D. P, Patterson, C. H, Yang, J. C, Henderson, Y. C, Yao, Y, Tour, J. M, Shashidhar, R, & Ratna, B. R. (2005). Molecularly inherent voltage-controlled conductance switching. Nature Materials, , 4

[29] Akkerman, H. B, Blom, P. W. M, De Leeuw, D. M, & De Boer, B. (2006). Towards molecular electronics with large-area molecular junctions. Nature, , 441

[30] Xu, y, & Musgrave, C. B. (2004). A DFT study of the Al_2O_3 atomic layer deposition on SAMs: effect of SAM termination. Chemistry of materials, , 16

[31] Kumar, A, Biebuyck, H. A, Abbott, N. L, & Whitesides, G. M. (1992). The use of self-assembled monolayers and a selective etch to generate patterned gold features. Journal of the American Chemical Society, , 114

[32] Kumar, A, & Whitesides, G. M. (1993). Features of gold having micrometer to centimeter dimensions can be formed through a combination of stamping with an elastomeric stamp and an alkanethiol "ink" followed by chemical etching. Applied Physics Letters, , 63

[33] Kumar, A, Biebuyck, H. A, & Whitesides, G. M. (1994). Patterning self-assembled monolayers: applications in materials science. Langmuir, , 10

[34] Carr, D. W, Lercel, M. J, Whelan, C. S, Craighead, H. G, Seshadri, K, & Allara, D. L. (1997). High-selectivity pattern transfer processes for self-assembled monolayer electron beam resists *Journal of Vacuum Science & Technology A*, , 15

[35] Huang, J. Y, & Hemminger, D. A. (1994). Photopatterning of self-assembled alkane-thiolate monolayers on gold: a simple monolayer photoresist utilizing aqueous chemistry. *Langmuir*, , 10

[36] Xu, S, & Liu, G. (1997). Nanometer-scale fabrication by simultaneous nanoshaving and molecular self-assembly *Langmuir*, , 13

[37] Ras, R. H. A, Sahramo, E, Malm, J, Raula, J, & Karppinen, M. (2008). Blocking the lateral film growth at the nanoscale in area-selective atomic layer deposition. *Journal of the American Chemical Society*, , 130

[38] Seo, E. K, Lee, J. W, Sung-suh, H. M, & Sung, M. M. (2004). Atomic layer deposition of titanium oxide on self-assembled-monolayer-coated gold. *Chemistry of materials*, , 16

[39] Sinha, A, Hess, D. W, & Henderson, C. L. (2006). Area-selective ALD of titanium dioxide using lithographically defined poly(methyl methacrylate) films. *Journal of The Electrochemical Society*, G465-G469, 153

[40] Färm, E, Kemell, M, Ritala, M, & Leskelä, M. (2008). Selective-area atomic layer deposition with microcontact printed self-assembled octadecyltrichlorosilane monolayers as mask layers. *Thin Solid Films* , 517

[41] Chen, R, & Bent, S. F. (2006). Chemistry for positive pattern transfer using area-selective atomic layer deposition. *Advanced Materials*, , 18

[42] Chen, R, & Bent, S. F. (2006). Highly stable monolayer resists for atomic layer deposition on Germanium and Silicon. *Chemistry of Materials* , 18

[43] Park, K, Lee, Y, Im, K. T, Lee, J. Y, & Lim, S. (2010). Atomic layer deposition of HfO_2 on self-assembled monolayer-passivated Ge surfaces. *Thin Solid Films* , 518

[44] Liu, J, Mao, Y, Lan, E, Banatao, D. R, Forse, G. J, Lu, J, Blom, H. O, Yeates, T. O, Dunn, B, & Chang, J. P. (2008). Generation of oxide nanopatterns by combining self-assembly of S-layer proteins and area-selective atomic layer deposition. *Journal of the American Chemical Society,*, 130

[45] Hong, J, Porter, D. W, Sreenivasan, R, Mcintyre, P. C, & Bent, S. F. (2008). ALD resist formed by vapor-deposited self-assembled monolayers. *Langmuir*, , 23

[46] Färm, E, Kemell, M, Ritala, M, & Leskelä, M. (2006). Self-assembled octadecyltrimethoxysilane monolayers enabling selective-area atomic layer deposition of iridium. *Thin Solid Films*, , 12

[47] Jiang, X, Huang, H, Prinz, F. B, & Bent, S. F. (2008). Application of atomic layer deposition of platinum to solid oxide fuel cells *Chemistry of materials*, , 20

[48]   Jiang, X, & Bent, S. F. (2007). Area-selective atomic layer deposition of platinum on YSZ substrates using microcontact printed SAMs. *Journal of The Electrochemical Society*, 154, DD656, 648.

[49]   Jiang, X, Chen, R, & Bent, S. F. (2007). Spatial control over atomic layer deposition using microcontact-printed resists. *Surface & Coatings Technology*, , 201

[50]   Ott, A. W, & Chang, R. P. H. (1999). Atomic layer-controlled growth of transparent conducting ZnO on plastic substrates. *Materials Chemistry and Physics*, , 58

[51]   Schoell, I. D, Sharp, S. J, Hoeb, M, Brandt, M. S, & Stutzmann, M. (2008). Electronic properties of self-assembled alkyl monolayers on Ge surfaces. *Applied Physics Letters*, , 92

[52]   Chen, R, Kim, H, Mcintyre, P. C, & Bent, S. F. (2004). Self-assembled monolayer resist for atomic layer deposition of HfO_2 and ZrO_2 high-κ gate dielectrics. *Applied Physics Letters*, , 84

[53]   Lao, S. X, Martin, R. M, & Chang, J. P. (2005). Plasma enhanced atomic layer deposition of HfO_2 and ZrO_2 high-k thin films. *Journal of Vacuum Science & Technology A*, , 23

[54]   Kim, H. (2003). Atomic layer deposition of metal and nitride thin films: Current research efforts and applications for semiconductor device processing. *Journal of Vacuum Science & Technology B*, , 21

[55]   Zhong, H, Heuss, G, & Misra, V. (2000). Electrical properties of RuO_2 gate electrodes for dual metal gate Si-CMOS. *Electron Device Letters, IEEE*, , 21

[56]   Hiratani, M, Nabatame, T, Matsui, Y, Imagawa, K, & Kimura, S. (2001). Platinum film growth by chemical vapor deposition based on autocatalytic oxidative decomposition. *Journal of The Electrochemical Society, C524-C527*, 148

[57]   Nayak, M, Ezhilvalavan, S, & Tseng, T. Y. (2001). High-Permittivity (Ba, Sr)TiO₃ thin films. Handbook of Thin Film Materials, Ed.: Nalwa, H. S.) Academic Press, San Diego, , 3, 99-167.

[58]   Wilk, G. D, Wallace, R. M, & Anthony, J. M. (2001). High-κ gate dielectrics: Current status and materials properties considerations. Journal of Applied Physics, , 89

[59]   Dasgupta, N. P, Lee, W, & Prinz, F. B. (2009). Atomic layer deposition of lead sulfide thin films for quantum confinement. *Chemistry of Materials*, , 21

[60]   Lee, J. P, Jang, Y. J, & Sung, M. M. (2003). Atomic layer deposition of TiO_2 thin films on mixed self-assembled monolayers studied as a function of surface free energy. *Advanced Functional Materials*, , 13

[61]   Choy, K. L. (2003). Chemical vapour deposition of coatings. *Progress in Materials Science*, , 48

[62]   Moshe, H, Vanbel, M, Valev, V, Verbiest, T, Dressler, D, & Mastai, Y. Chiral nanosized metal oxide surfaces. unpublished yet.

[63] Chena, Q, & Richardsonb, N. V. (2004). Physical studies of chiral surfaces. *Annual Reports Section "C"*, , 100

[64] Shval, A, & Mastai, Y. (2011). Isothermal titration calorimetry as a new tool to investigate chiral interactions at crystal surfaces. *Chemical Communications*, , 47

Chapter 3

Plasma Electrolytic Oxidation of Valve Metals

Alex Lugovskoy and Michael Zinigrad

Additional information is available at the end of the chapter

1. Introduction

Plasma electrolytic oxidation (PEO) is also known as micro-arc oxidation and spark anodizing is often regarded as a version of anodizing of valve metals (Mg, Al, Ti, and several others) and their alloys. Indeed, the essence of both anodizing and PEO is the production of oxide layers on a metal surface by the action of electricity in a convenient electrolyte. An oxide layer has a complex composition and includes various oxides of a base metal, alloy additives and species coming from the electrolyte. For both anodizing and PEO, an oxide layer forms due to electrochemical oxidation of the metal constituents and inclusion of some components of the electrolyte with possible further interactions in the vicinity of the electrode.

However, some features of PEO are clearly different than those of the anodizing. Normally, low-voltage direct currents are used for anodizing and the formation of the oxide layer occurs under a quiescent regime. The produced layer has relatively homogeneous structure with more or less evenly distributed blind pores. The thickness of an oxide layer is limited by ~ 20-50 μm for most cases, because the electrical conductivity of the oxide layer is low and the formation of the layer effectively breaks the electric circuit in the cell. It can be said that the formed layer "passivates" the metal surface in the course of anodizing and thus prevents its own further growth (Fig. 1, stages I - II). It deserves to be notes that sparking is considered undesirable for anodizing, because it is an indication of cracks and inhomogeneities in the formed layer.

For PEO, significantly higher voltages and (normally) alternating currents cause intense sparking due to micro-arc discharges that break down the oxide layer (Fig.1, stages I – III). Extreme temperatures and pressures [1] develop in the discharge channels and cause complex phase-transformation processes that result in the production of a compact, thick hard layer, which often has attractive abrasion and corrosion resistances. Although PEO layers have a relatively high porosity [2], they can effectively protect the base metal against corrosion because the pores formed by a discharge can subsequently "heal" by molten oxides due to high local temperatures in the vicinity of plasma discharge channels [3] and are therefore impermeable to corrosion media. The improved corrosion stability of PEO-treated metals as compared to bare metals has been reported for aluminum [4-8] and magnesium alloys (see, for instance, [9, 10]).

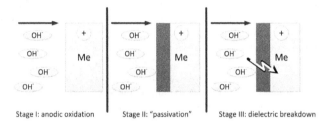

Stage I: anodic oxidation   Stage II: "passivation"   Stage III: dielectric breakdown

Figure 1. Main stages of an oxide layer formation in the processes of anodizing (stages I - II) and PEO (stages I – III)

Both direct and alternating current can be used for PEO. However, AC regime is preferable, because pores formed during a cathodic breakdown "heal" by molten oxides during the next anodic pulse [3], the electrolyte in the metal vicinity is refreshed and the produced oxide layers are more uniform. Industrial 50–60 Hz sine-wave AC voltages of 100–600V are most frequently used for the PEO processing. Due to the partial rectifying effect of the valve metal oxide, complex sew-like waves are observed in practice (Fig. 2).

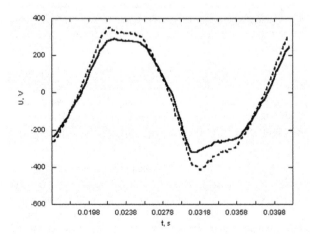

Figure 2. Observed voltage profile of a typical PEO processing of an aluminium alloy in alkaline silicate electrolytes containing 10 g/L of $Na_2O\ SiO_2$ (solid line) or of $Na_2O\ 3SiO_2$ (dotted line). Industrial 50 Hz sine-wave voltage of 200V was supplied for both cases.

The detailed mechanism of the PEO process has not yet been revealed; however, most investigators (see, for example, [3, 11]) agree that during each AC period several principal steps occur: (1) a barrier oxide layer forms on the boundary between the metal and the electrolyte during the initial anodic semi-period; (2) the potential difference between the two sides of the dielectric oxide layer increases as the anodic semi-period advances until (3) dielectric break-down takes place. The breakdowns through the oxide layer are accompanied by sparks, so that the process actually occurs in a mode of micro-arc discharges. Fresh portions of the electrolyte are injected into the bare metal surface during the breakdowns, and the process continues as long as the voltage is sufficient for new breakdowns which perforate the growing oxide layer. Relaxation of the metal and oxide and partial reduction of the oxidized species occur during cathodic semi-periods. Gas micro-phase formation (nucleation) and annihilation (cavitation) processes apparently contribute to formation of the oxide layer, but these processes have been scarcely studied because of obvious experimental difficulties.

The production of oxide layers by PEO was most often studied for aluminum and its alloys (for example, in [5-8], [12-17]), magnesium and its alloys ([9, 10] and others), titanium and its alloys (for example, in [18-19]). Studies of PEO on other metals (zirconium [20], zirconium alloys [21] and steel [22]) are scarce.

The various versions of PEO differ from each other with respect to the profile of the applied voltage and the composition of electrolyte. The oxide layer produced always consists of two sub-layers: an outer brittle sub-layer, which typically has a hardness of 500–1000 HV and a porosity of more than 15%, and an inner functional sub-layer, whose typical hardness is 900–2000 HV and whose typical porosity is 2–10%. The outer brittle sub-layer can be easily removed by polishing, and the inner harder sub-layer can be finished to a smooth marble surface.

Oxide layers can be produced in several types of electrolytes, whose action can differ (see the survey in [11]). Acidic and alkaline electrolytes dissolve moderately the base metal, phosphate and polymer electrolytes passivate it, and fluoride electrolytes interact with it in more complex and less understood ways.

The most frequently used electrolytes for the PEO processing of aluminum and aluminum alloys are aluminate [23, 24], phosphate [24, 25] and, most often, alkaline silicate solutions (for example, [7, 11]). Magnesium and its alloys are normally PEO processed in alkaline phosphate [26-28] or alkaline silicate electrolytes [10, 29, 30] often containing fluorides [27-30]. Aluminate, phosphate and silicate electrolytes are used for titanium and its alloys (cf. a comparative study in [31]).

Since silicate electrolytes are frequently used for the PEO processing of the valve metals, many aspects of their influence on the properties of produced oxide layers have been intensively studied. For example, it was established that the addition of silicates to the electrolytes stabilizes the oxide layer toward alkaline attacks [12], causes some increase in the thickness of the oxide layers, but reduces their hardness and wear resistance as compared to alkaline electrolytes without silicates [13].

Less information is available about the effect of various forms of "water glasses," i.e., polymer silicates of various composition, on the structure and properties of PEO layers. Little or nothing is known about the difference between oxide layers obtained in silicate electrolytes having identical or close element composition, but containing silicates of different SiO_2-to-Na_2O ratios (silicate indexes).

Another point of interest is the influence of the fluoride additives on the structure and properties of the PEO oxide layers produced not on aluminum alloys only, but also on other base metals.

Here we try to summarize these two effects (the role of the silicate index and the influence of fluorides) in a comparative study of the PEO processing of a magnesium alloy and of an aluminum alloy.

2. Experimental

Rectangular flat (3 x 15 x 30 mm) specimens of aluminum A5052 alloy (Al as the base and approximately 2.5% of Mg) and magnesium AZ9110D alloy (Mg as the base and 8.3-9.7% Al, 0.15% Mn min., 0.35-1.0% Zn, 0.10% Si max., 0.005% Fe max., 0.030% Cu max., 0.002% Ni max., 0.02% max. others) were cut, polished with #1200 grit SiC abrasive paper and rinsed in tap water prior to be PEO processed. The oxidation was performed in AC mode by the industrial 50 Hz sine voltage at the end current density 6.6 ± 0.2 A / dm^2 for 30-60 minutes on a home-made 40 kVA PEO station with a water-cooled bath made of stainless steel, which served as the counter electrode. Potassium hydroxide KOH (Finkelman Chemicals, technical grade), KF (Merck, 99%), sodium silicate Na_2O SiO_2 $5H_2O$ (pentahydrate, Spectrum, practical grade), and water glass Na_2O $3SiO_2$ (Spectrum, practical grade) having the silicate indexes n = 1 and n=3, respectively were used for the preparation of the electrolytes.

Conductivities and pH of the electrolytes were measured by a YK-2005WA pH/CD meter, the thickness of oxide layers was first roughly measured by a coating thickness gauge CM-8825 and then more exactly by SEM. The surface morphology, structure and composition were inspected on SEM JEOL JSM6510LV equipped with an NSS7 EDS analyzer (Correction Method Proza – Phi-Pho-Z was used for the quantitative analysis). Cross-section samples prepared according to standard metallographic protocols [32] were used for SEM, EDS, XRD and microhardness measurements. Microhardness was measured on Buehler Micromet 2100, HV_{25}. X-ray Diffractometer (XRD) Panalytical X'Pert Pro with Cu $K\alpha$ radiation (λ=0.154 nm) was used with the full pattern identification made by X'Pert HighScore Plus software package, version 2.2e (2.2.5) by PANalytical B.V. Materials identification and analysis made by the PDF-2 Release 2009 (Powder Diffraction File). Phase analysis identification made by XRD, 40kV, 40mA. The XRD patterns were recorded in the GIXD geometry at a=1°and 5° in the range of 20-80º (step size 0.05º and time per step 2s).

Autolab12 Potentiostat with a standard corrosion cell was used for corrosion tests. Potentials were measured against Ag | AgCl reference electrode and then related to SHE.

3. Results and discussion

3.1. PEO of aluminum A5052 alloy in different alkaline silicate electrolytes

Two sodium silicates were taken for the comparison. The first, $Na_2O\ SiO_2$ will be hereafter referred to as the "n=1 silicate" and the second, $Na_2O\ 3SiO_2$ will be referred to as the "n=3 silicate," in accordance to their silicate index, that their SiO_2-to-Na_2O ratios. PEO processing was performed in the electrolytes containing 1 gr/L (17.9 mmol/L) KOH and various amounts of the silicates as specified in Table 1. Conductivities of the electrolytes were at least 4-5 mS/m and all the electrolytes had pH = 11-13 (see Table 1).

$Na_2O\bullet nSiO_2$	5 g/L	10 g/L	15 g/L
$Na_2O\cdot SiO_2$ (n = 1)	12.68 / 10.27	12.74 / 15.5	12.80 / 22.7
molarity	0.021	0.041	0.062
$Na_2O\cdot 3SiO_2$ (n = 3)	11.08 / 4.53	11.18 / 5.47	11.24 / 6.52
molarity	0.024	0.047	0.071

Table 1. Typical electrolyte parameters (pH / Conductivity, mS/m)

As seen from Table 1, both the basicity and the conductivity are strongly affected by the silicate index, which is not surprising because the molar fraction of sodium oxide is 0.5 for Na_2O SiO_2 and only 0.25 for $Na_2O\ 3SiO_2$. The values of pH of the electrolytes only weakly depend on the concentration of a given silicate, while their conductivities are roughly proportional to the concentration of $Na_2O\ SiO_2$ or $Na_2O\ 3SiO_2$. As one could expect, better conductivities of the "n=1 electrolytes" must facilitate the PEO process.

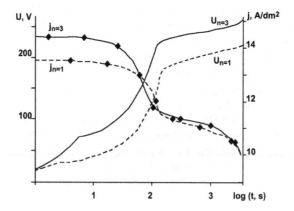

Figure 3. Typical voltage (U) and current density (j) amplitudes in the PEO process of an aluminium alloy in n=1 and n=3 electrolytes.

Indeed, lower current densities are needed for the plasma process initiation when $n = 1$ (Fig. 3). As seen from Fig. 3, not only the initial current densities, but also the process voltages are higher for "n = 3 electrolytes". Visual changes both in voltage and in current density are observed after 100 - 200 seconds. These changes can indicate that a steady state has been achieved, when only significantly fewer discharges occur and the oxide layer has mainly been formed.

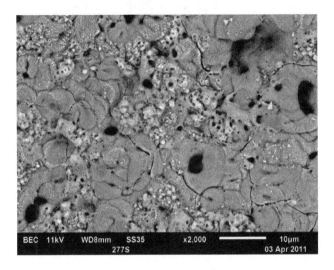

Figure 4. A SEM image (Secondary electron detector) of the morphology of an oxide layer produced by PEO of an aluminium alloy in an alkaline silicate electrolyte.

Oxide layers produced after 30-60 min of PEO have porous morphology with blind "crater-like" pores, which are the results of plasma discharges through the oxide (Fig. 4). No difference in morphology was observed for the two types of electrolytes.

The morphology and elemental composition of a pore obtained by the EDS are presented in Fig. 5 and Table 2. As follows from the data of the elemental analysis, the interior of a pore contains much less silicon and much more aluminum than the exterior close to the surface. This is not surprising, because aluminum comes from inside (from the metal substrate), while silicon is provided by the electrolyte and only with difficulties can penetrate to the depths of the oxide layer.

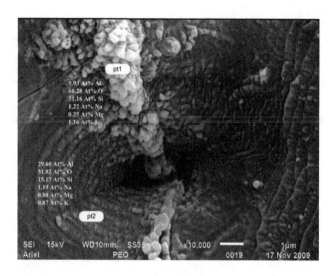

Figure 5. Morphology and elemental composition of a pore at two characteristic points: point 1 (pt1) on the surface of the specimen; point 2 (pt2) inside the pore.

At.%	Al	Si	O	Na	Mg	K
Point 1	5.9	31.2	60.3	1.2	0.3	1.2
Point 2	29.6	15.2	51.8	1.2	0.8	0.9

Table 2. Element composition of the oxide coating on the surface (point 1) and inside a pore (point 2)

The oxide layers, formed after 30 minutes of PEO, are 20-60 μm thick and consist of two clearly pronounced sublayers: a denser inner sublayer and a loose porous outer sublayer (Fig. 6).

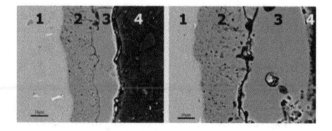

Figure 6. The structure of oxide layers on cross-sections of specimens obtained by PEO in (left image) 0.05 mol/L n=1 and (right image) n=3 electrolytes: (1) non-oxidized base alloy, (2) inner denser oxidized sublayer, (3) outer loose sublayer, (4) resin wrapping. Back-Scattered Electron SEM image, x1,000.

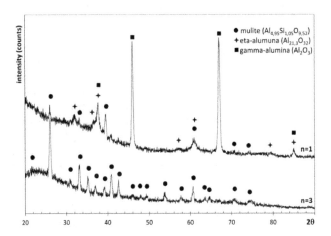

Figure 7. XRD patterns for oxide layers on specimens obtained by PEO in 0.05 mol/L n=1 and n=3 electrolytes. Minor phases are not shown.

While the outer sublayer contains fewer pores, the pores in it are much larger than in the inner sublayer. The mean density of the outer sublayer is lower than of the inner sublayer. Both sublayers are considerably thicker for the "n=3 electrolytes", but they contain 2-5 times as much silicon as for the "n=1 electrolytes". XRD phase analysis (Fig. 7) shows that the oxide layer contains 60-70% of γ-alumina, 20-30% of η-alumina and about 1% of quartz for the "n=1 electrolytes." For "n=3 electrolytes" the oxide layer consists mainly of mullite $3Al_2O_3 2SiO_2$ and varied (for different concentrations of the n=3 silicate in the electrolyte) amounts of amorphous silica, quartz and various types of alumina.

Silicate concentration in electrolyte, mol·L⁻¹	Microhardness, Vickers (HV)			
	n=1 electrolyte, outer sublayer	n=1 electrolyte, inner sublayer	n=3 electrolyte, outer sublayer	n=3 electrolyte, inner sublayer
0.019	840	1100	770	1060
0.025	1130	1380	1280	1570
0.050	890	1050	710	980
0.075	920	1630	700	910
Mean hardness	945	1290	865	1130

Table 3. Microhardness of oxidized sublayers produced by PEO in electrolytes containing different concentrations of "n=1" and "n=3" silicates measured on cross-sectioned specimens perpendicularly to the section planes.)

Results of hardness measurements of oxide layers are presented in Table 3. Obviously, alumina-quartz layer formed in the "n=1 electrolytes" is harder than mullite layer formed in the "n=3 electrolytes."

Corrosion tests were made after a specimen was masked by resin except for a square window having the area of 1 cm² on the oxidized surface. Thus prepared specimen was held for 1 hour in 1% NaCl for the achievement of steady-state corrosion and then its voltammetric curve was measured using Linear Sweep Voltammetry (25 mV/sec). Broader potential range (normally, OCP ± 500 mV) was first studied for the determination of the corrosion potential. Narrower potential range of ± 50-70 mV relatively to the previously roughly determined corrosion potential was then measured and used for Tafel slope analysis. All potentials were measured against the Ag|AgCl reference electrode and then recalculated to the standard hydrogen electrode potentials.

The results of thus measured corrosion characteristics of "bare" Al5052 alloy and different PEO oxidized specimens are given in Table 4. As follows from Table 4, corrosion current densities measured on oxidized samples are at least 3-4 times lower than for the untreated alloy. Corrosion potentials for all the oxidized samples are considerably more positive than for the untreated alloy, which evidences the increase of anodic stability in the test solution. The most noble corrosion potentials are observed for lower concentrations of both n=1 and n=3 silicates and correlate with higher microhardness of oxide layers (Table 3) observed for these concentrations. We could carefully assume that the content of γ-alumina in an oxide layer plays the key role in the shift of corrosion potentials to the positive direction.

Silicate concentration in electrolyte, mol·L⁻¹	n=1 electrolyte, E_{corr}, V vs. SHE	n=1 electrolyte, i_{corr}, A/cm² x 10⁶	n=3 electrolyte, E_{corr}, V vs. SHE	n=3 electrolyte, i_{corr}, A/cm² x 10⁶
"bare" Al5052	-1.126	15.99		
0.013			-0.525	0.08
0.025	-0.497	3.60	-0.815	2.66
0.050	-0.796	4.30	-0.998	0.98
0.075			-0.972	1.68
0.100	-0.942	1.93		
0.150	-0.995	3.77		

Table 4. Corrosion current densities and potentials of Al5052 alloy oxidized in different electrolytes.

The results summarized in Table 4 are better than those obtained for anodizing [14], similar to those obtained for much more expensive protection methods and similar or batter than those obtained by PEO in other silicate electrolytes [6, 15-17]. All the measurements evidence that specimens treated in "n=3 electrolytes" have better corrosion protection than those treated in "n=1 electrolytes." The microscopic inspection of cross-sections evidences (Fig. 4) that even though the outer sublayer produced in "n=3 electrolytes" contains large caverns and the inner sublayer looks more porous than for the "n=1 electrolytes", the larger thickness of the layer produced in the "n=3 electrolytes" presents a more difficult barrier for the diffusion of corrosive media and therefore forms better protection of the metal substrate. It deserves to be reminded that the oxide layers produced in the "n=1 electrolytes" only contain oxide phases (alumina and quartz), in contrast to the mullite layer produced in the "n=3 electrolytes." Obviously, the milder mullite better fills pores in the oxide layer than harder oxides do. Somewhat similar results were obtained by another research group [7] for another Al alloy (2219) and also demonstrated that higher silicate contents in silicate-alkaline electrolytes increase the corrosion resistance of PEO coatings.

3.2. Fluoride influence on the properties of oxide layer produced by PEO

The surface of PEO layers produced on both the magnesium and the aluminum alloys is normal for PEO coatings and consists of "volcanic" pores chaotically distributed on a fused surface (Fig. 4).

Long PEO processing times (30-90 min) result in thick coatings for both alloys, but the initial stages demonstrate a clear difference between the two metal alloys. For the PEO treatment in the alkaline silicate electrolyte (0.08 mol /L KOH + 0.08 mol / L Na_2SiO_3) without the fluoride addition, the 15 minute process produces approximately 20 μm thick non-continuous oxide layer on the magnesium alloy and only about 5 μm thick non-continuous layer on the aluminum alloy (Fig. 8).

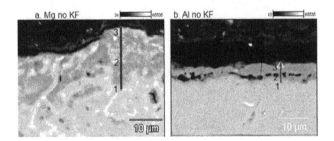

Figure 8. SEM images (x3000) with EDS linear scans of cross-sectioned PEO layers on (a) the magnesium and (b) aluminum alloys obtained after 15 minute oxidation. The black zone in the topmost parts is polymer tar fixing the specimen. The base metals are in the bottom part of each image. The element composition at points 1, 2, 3 of each image is specified in Table 5.

Both on the magnesium and on the aluminum alloys the coating is porous, but the porous are blind. Element compositions at points 1, 2 and 3 along lines drawn from the pure base metal to the outer border of the coatings (Fig. 8) are given in Table 5.

	Mg alloy		Al alloy
Point 1	Mg 82 at%, Al 8 at%, O 9 at%, Si 1 at%	Point 1	Al 94 at%, Mg 2 at%, O 4 at%
Point 2	Mg 67 at%, Al 13 at%, O 17 at%, Si 3 at%	Point 2	Al 61 at%, Mg 3.5 at%, O 34 at%, Si 1.5 at%
Point 3	Mg 60 at%, Al 9 at%, O 27 at%, Si 4 at%	Point 3	Al 70 at%, Mg 2.5 at%, O 15 at%, Si 9.5 at%, K 3 at%

Table 5. Element compositions at representative points 1, 2 and 3 as shown in Fig. 8.

As follows from Fig. 8 and Table 5, the oxidized layer on the magnesium alloy is not only thicker, but also more uniform than that on the aluminum alloy, for which large voids containing relatively high amounts of oxygen are formed between the base metal and the oxide layer. It deserves to be noted that the percentage of oxygen is lower than what should be expected according to the stoichiometry of magnesium and aluminum oxides (Mg : O = 1:1 and Al : O = 2 : 3). Therefore, the oxide layer never consists of the oxides only, but contains some excess metal atoms.

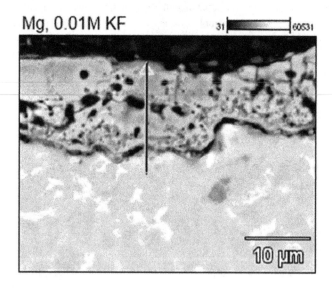

Figure 9. BSE SEM images (x3000) with EDS linear scans of cross-sectioned PEO coating on the magnesium alloy obtained after 15 minute oxidation in the electrolyte containing 0.01 mol/ L KF.

As KF has been added to the electrolytes, the situation with the oxidation of magnesium alloy becomes different. While the total thickness of the coating remains relatively the same (~ 20 μm after 15 minutes), its structure and composition are clearly distinguished from what was observed without the fluoride. Already for the lowest studied KF concentration (0.01 mol / L) the coating is continuous and consists of two very different sublayers (Fig. 9). The inner sublayer is only 2-3 μm thick and contains about 70 at% of Mg, 25 at% of O and 5 at% of Al. The outer sublayer is 15-18 μm thick, porous and contains large voids filled by light elements (darker sites in Fig. 9). The typical composition of the outer layer is 50-60 at% O, 35-40 at% Mg, 10 at% Si and 2-4 at% Al. As the content of KF in the electrolyte increases, the structure of the layers does not change, but significant amounts of fluorine are detected in the coating.

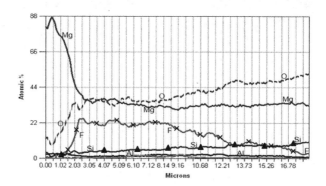

Figure 10. Element composition profiles (EDS linear scan) across the PEO coating on the magnesium alloy obtained after 15 minute oxidation in the electrolyte containing 0.2 mol/ L KF.

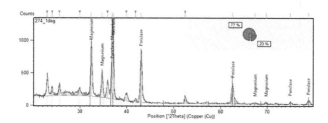

Figure 11. Small angle (1º) XRD pattern for the surface of the PEO coating on the magnesium alloy obtained after 15 minute oxidation in the electrolyte containing 0.02 mol/ L KF.

Interestingly, the maximal amount of fluoride was detected in the most inward part of the outer sub-layer (Fig. 10). According to small angles (1º-5º) XRD measurements, the surface consists of Periclase MgO (77%) and metal Mg (23%). Deeper layers of the coating demonstrate the increase of Mg at the expense of MgO. No fluorine-containing phases could be identified with confidence (Fig. 11).

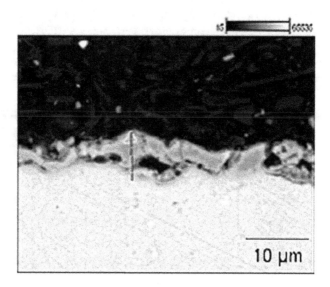

Figure 12. BSE SEM images (x3000) with EDS linear scans of cross-sectioned PEO coating on the aluminum alloy obtained after 15 minute oxidation in the electrolyte containing 0.2 mol/ L KF.

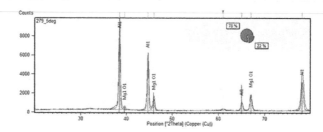

Figure 13. Small angle (5º) XRD pattern for the surface of the PEO coating on the aluminum alloy obtained after 15 minute oxidation in the electrolyte containing 0.05 mol/ L KF.

The addition of KF to the electrolytes for the PEO processing of the aluminum alloy has another effect. While for lower concentrations of KF (<0.05 mol / L) only a very thin porous coating layer is formed, higher KF contents result in the formation of 5-10 µm oxide layer. As the thickness of the coating increases, two sublayers are revealed. As for the magnesium alloy, the thin inner sublayer is denser and the thicker outer one is more porous (Fig. 12). As evidenced by EDS, the outer sublayer contains 60 at% of O and 40 at% of Al. As much as 9% fluorine is found in the inward part of the outer sublayer. XRD measurements show that the surface (1º incident beam) consists of two crystal phases, whose composition is Al_2MgO_4 (56%) and MgO

(44%). Deeper layers (5º incident beam) are formed by metal Al (78%) and MgO (22%). A sample of an XRD pattern is given in Fig. 13.

The most obvious observation, which can be made from the comparison of the PEO of the magnesium and of the aluminum alloys, is that magnesium is oxidized much more easily than aluminum. While for all the studied magnesium systems 20 μm coating was produced after 15 minutes of the PEO, only 5-10 μm coating on the aluminum alloy could be obtained for the same process time. An interesting issue is the ratio "Mg:Al" at different depth of the coatings (Table 6).

	C_{KF}, mol / L	Base metal	Inner sublayer	Outer sublayer
	0	10.3	6.7*	
Mg alloy	0.01	10.3	14	8.2
	0.1	10.3	15	6.4
	0	0.02	0.06	0.04
Al alloy	0.01	0.02	0.04	0.06
	0.1	0.02	0.04	0.05

*No sublayers were observed for this PEO layer.

Table 6. "at% Mg : at% Al" ratio at different depths.

As follows from Table 6, magnesium content at the inner sublayer is always higher than in the base metal, except for the oxidation of the magnesium alloy in the electrolyte containing no fluoride, when no sublayers can be seen. The trend is less straightforward for the outer sublayer, which can be explained by the fact that the latter is thicker, more porous and much less uniform. XRD phase analysis confirms than the key role at the initial stage of the oxidation is played by magnesium oxide and only on the surface aluminum starts to be oxidized to form Al_2MgO_4. These can be explained by two facts: (1) that the amphoteric aluminum is readily dissolved by the alkaline electrolyte while magnesium is not; (2) that according to Ellingham diagrams [33] the oxidation of magnesium is thermodynamically more favorable than that for aluminum in the entire range of temperatures below 1500K.

Many authors report that they could not identify a fluoride containing phase in PEO coatings on aluminum or magnesium alloys obtained in alkaline fluoride-containing electrolytes (see, for instance [34-37]). This is consistent with our XRD observation; however we could clearly see the presence of fluoride on EDS cross-sectional profiles of the coatings (Fig. 4). This means that considerable amounts of amorphous fluorides are found in the coatings very close to the base metal. Summarizing, the action of fluoride additives can be assumed as follows: (1) fluoride anions are first chemisorbed to the metal surface and create on it barrier layer; (2) as the dielectric breakdown occurs, oxide layer is formed due to the exchange of fluorine by oxygen atoms from the electrolyte; (3) fluoride remains in the vicinity of the metal and takes part in the formation of an amorphous phase. This process is much more favorable for the magnesium alloy due to the easier breakdown of magnesium fluoride as compared to aluminum fluoride.

4. Conclusions

Plasma Electrolytic Oxidation in alkaline silicate electrolytes containing 0.013-0.150 mol/L of sodium silicates having silicate index n=1 or n=3, was performed on Al5052 aluminum alloy. For all the electrolytes studied, 20-90 μm thick oxide layer was obtained and its composition, structure and properties were studied. For each sample, the oxidized layer consists of a denser inner and looser outer sublayer. While for "n=1 electrolytes" the oxidized layer is mainly formed by several kinds of alumina, the principal constituent of the oxidized layer for "n=3 electrolytes" is mullite.

Measurements of microhardness evidenced that it is apparently not influenced by the kind of silicate (n=1 or n=3) and by its concentration in the electrolyte.

Electrolytes with silicate index n=3 ensure better corrosion protection than those with n=1. This might be caused by the milder and more plastic nature of the oxide layer produced in the "n=3 electrolutes" as compared to those produced in the "n=1 electrolytes."

Corrosion protection parameters are significantly better for all PEO oxidized samples than for the untreated Al5052 alloy.

The formation of PEO coating on magnesium and aluminum alloys in the presence of fluoride starts with the fluorination of the metal surface and formation of a dielectric metal fluoride layer. Electric breakdowns destruct this layer and form oxide layers containing also amorphous fluorides.

The fluoride-supported PEO process proceeds more easily for the magnesium than for the aluminum alloys. This difference might be caused by the easier breakdown of the dielectric layer containing magnesium fluoride as compared to that containing aluminum fluoride. This is consistent with the values of dielectric permeability of magnesium fluoride (4.87) [38] and of aluminum fluoride (2.2) [39].

The external surface of the coating is enriched by magnesium as compared to the base metal.

Author details

Alex Lugovskoy* and Michael Zinigrad

*Address all correspondence to: lugovsa@ariel.ac.il

Chemical Engineering Department, Ariel University Center of Samaria, Ariel, Israel

References

[1] Hussein, R. O, Nie, X, Northwood, D. O, Yerokhin, A, & Matthews, A. Spectroscopic study of electrolytic plasma and discharging behaviour during the plasma electrolytic oxidation (PEO) process. J. Phys. D: Appl. Phys. (2010).

[2] Curran, J. A, & Clyne, T. W. Porosity in plasma electrolytic oxide coatings. Acta Materialia (2006)., 54-1985.

[3] Sah, S. P, Tsuji, E, Aoki, Y, & Habazaki, H. Cathodic pulse breakdown of anodic films on aluminium in alkaline silicate electrolyte- Understanding the role of cathodic half-cycle in AC plasma electrolytic oxidation. Corrosion Science (2012)., 55-90.

[4] Barik, R. C, Wharton, J. A, Wood, R. J. K, Stokes, K. R, & Jones, R. L. Corrosion, erosion and erosion-corrosion performance of plasma electrolytic oxidation (PEO) deposited Al₂O₃ coatings. Surface & Coatings Technology (2005)., 199-158.

[5] Xue, W, Shi, X, Hua, M, & Li, Y. Preparation of anti-corrosion films by microarc oxidation on an Al-Si alloy. Applied Surface Science (2007)., 253-6118.

[6] Nie, X, Meletis, E. I, Jiang, J. C, Leyland, A, Yerokhin, A. L, & Matthews, A. Abrasive weary corrosion properties and TEM analysis of Al₂O₃ coatings fabricated using plasma electrolysis. Surface and Coatings Technology (2002)., 149-245.

[7] Alsrayheen, E, Campbell, B, Mcleod, E, Rateick, R, & Birss, V. Exploring the effect of alkaline silicate solution composition on the ac/dc spark anodization of Al-Cu Alloys. Electrochimica Acta (2012)., 60-102.

[8] Venugopal, A, Panda, R, Manwatkar, S, & Sreekumar, K. Rama Krishna L., Sundararajan G. Effect of micro arc oxidation treatment on localized corrosion behaviour of AA7075 aluminum alloy in 3.5% NaCl solution. Trans. Nonferrous Met. Soc. China (2012).

[9] Guo, H, & An, M. Xu Sh., Huo H., Microarc oxidation of corrosion resistant ceramic coating on a magnesium alloy. Materials Letters (2006)., 60-1538.

[10] Chen, F, Zhou, H, & Yao, B. Qin Zh., Zhang Q. Corrosion resistance property of the ceramic coating obtained through microarc oxidation on the AZ31 magnesium alloy surfaces. Surface & Coatings Technology (2007)., 201-4905.

[11] Yerokhin, A. L, Nie, X, Leyland, A, Matthews, A, & Dowey, S. J. Plasma electrolysis for surface engineering. Surface and Coatings Technology (1999)., 122-73.

[12] Moon, S, & Jeong, Y. Generation mechanism of microdischarges during plasma electrolytic oxidation of Al in aqueous solutions, Corrosion Science (2009)., 51(2009), 1506-1512.

[13] Polat, A, Makaracib, M, & Ustac, M. Influence of sodium silicate concentration on structural and tribological properties of microarc oxidation coatings on 2017A aluminum alloy substrate. Journal of Alloys and Compounds (2010)., 504-519.

[14] Zabielski, C. V, & Levy, M. Study of Type II and Type III Anodized Al in Aqueous DS2 Solutions. U.S. Army Research Laboratory Environmental Effects. In: Proceedings of the TRI-Service conference on corrosion. Plymouth, Massachusetts 12-14 May (1992). Avavilable on http://namis.alionscience.com/conf/tscc/search/pdf/AM026095.pdf., 5052-0.

[15] Tseng ChCh., Lee J.-L., Kuo Tz.-H., Kuo Sh.-N., Tseng K.-H. The influence of sodium tungstate concentration and anodizing conditions on microarc oxidation (MAO) coatings for aluminum alloy. Surface & Coatings Technology (2012). , 206-3437.

[16] Raj, V. Mubarak Ali M. Formation of ceramic alumina nanocomposite coatings on aluminium for enhanced corrosion resistance. J. Mat. Proc. Technology (2009). , 209-5341.

[17] Wei, C. B, Tian, X. B, Yang, S. Q, Wang, X. B, Fu, R. K. Y, & Chu, P. K. Anode current effects in plasma electrolytic oxidation. Surface & Coatings Technology (2007). , 201-5021.

[18] Li, Y, Yao, B, Long, B. Y, Tian, H. W, & Wang, B. Preparation, characterization and mechanical properties of microarc oxidation coating formed on titanium in $Al(OH)_3$ colloidal solution. Applied Surface Science (2012). , 258-5238.

[19] Wang, Y. M, Jiang, B. L, Lei, T. Q, & Guo, L. X. Microarc oxidation coatings formed on Ti_6Al_4V in Na_2SiO_3 system solution: Microstructure, mechanical and tribological properties. Surface & Coatings Technology (2006). , 201-82.

[20] Pauporté, T, Finne, J, Kahn-harari, A, & Lincot, D. Growth by plasma electrolysis of zirconium oxide films in the micrometer range. Surface & Coatings Technology (2005). , 199-213.

[21] Cheng, Y, Matykina, E, Arrabal, R, Skeldon, P, & Thompson, G. E. Plasma electrolytic oxidation and corrosion protection of Zircaloy-4. Surface & Coatings Technology (2012). , 206-3230.

[22] Wang, Y. Jiang Zh., Yao Zh. Preparation and properties of ceramic coating on Q235 carbon steel by plasma electrolytic oxidation. Current Applied Physics (2009). , 9-1067.

[23] Xin ShG., Song L.-X., Zhao R.-G., Hu X.-F. Properties of aluminium oxide coating on aluminium alloy produced by micro-arc oxidation. Surface & Coatings Technology (2005). , 199-184.

[24] Shen, D, & Wang, J. Y.-L., Nash Ph., Xing G.-Zh. Microstructure, temperature estimation and thermal shock resistance of PEO ceramic coatings on aluminium. J. materials processing technology (2008). , 205-477.

[25] Snizhko, L. O, Yerokhin, A. L, Pilkington, A, Gurevina, N. L, Misnyankin, D. O, Leyland, A, & Matthews, A. Anodic processes in plasma electrolytic oxidation of aluminium in alkaline solutions. Electrochimica Acta (2004). , 49-2085.

[26] Timoshenko, A. V. Magurova Yu.V. Investigation of plasma electrolytic oxidation processes of magnesium alloy MAunder pulse polarisation modes. Surface & Coatings Technology (2005). , 2-1.

[27] Boinet, M, Verdier, S, Maximovitch, S, & Dalard, F. Plasma electrolytic oxidation of AM60 magnesium alloy: Monitoring by acoustic emission technique. Electrochemical properties of coatings. Surface & Coatings Technology (2005). , 199-141.

[28] Hsiao, H, Tsung, Y, Ch, H, Tsai, W, Anodization, T, & Of, A. Z. D magnesium alloy in silicate-containing electrolytes. Surface & Coatings Technology (2005). , 199-127.

[29] Duan, H. Yan Ch., Wang F. Effect of electrolyte additives on performance of plasma electrolytic oxidation films formed on magnesium alloy AZ91D. Electrochimica Acta (2007). , 52-3785.

[30] Liang, J, Guo, B, Tian, J, Liu, H, Zhou, J, & Xu, T. Effect of potassium fluoride in electrolytic solution on the structure and properties of microarc oxidation coatings on magnesium alloy. Applied Surface Science (2005). , 252-345.

[31] Yerokhin, A. L, Nie, X, Leyland, A, & Matthews, A. Characterisation of oxide films produced by plasma electrolytic oxidation of a Ti-6Al-4V alloy. Surface & Coatings Technology (2000). , 130-195.

[32] Vander Voort GF., editor. ASM Handbook, and Microstructures. (2004). ASM International., 09-Metallography

[33] MIT educational materialshttp://web.mit.edu/2.813/www/readings/Ellingham_diagrams.pdfaccessed on August 31, (2012).

[34] Wang ZhWu L., Cai W., Shan A, Jiang Zh. J. Effects of fluoride on the structure and properties of microarc oxidation coating on aluminium alloy. Alloys and Compounds (2010). , 505-188.

[35] Wang, K, Koo, B, Lee, H, Ch, G, Kim, Y, Lee, J, & Byon, S. -H. E. Effects of electrolytes variation on formation of oxide layers of 6061 Al alloys by plasma electrolytic oxidation. Trans. Nonferrous Met. Soc. China (2009). , 19-866.

[36] Gnedenkov, S. V, Khrisanfova, O. A, Zavidnaya, A. G, Sinebrukhov, S. L, Gordienko, P. S, Iwatsubo, S, & Matsui, A. Composition and adhesion of protective coatings on aluminium. Surface and Coatings Technology (2001). , 145-146.

[37] Lin, C. S, & Fu, Y. C. J. Characterization of anodic films of AZ31 magnesium alloys in alkaline solutions containing fluoride and phosphate anions. J. Electrochem. Soc. (2006). BB424., 417.

[38] Duncanson, A, & Stevenson, R. W. H. Some Properties of Magnesium Fluoride crystallized from the Melt. Proc.Phys.Soc. (1958). , 72, 1001.

[39] Table of Dielectric ConstantsASI Instruments Web Site. http://wwwasiinstr.com/technical/Dielectric%20Constants.htmaccessed on November 3, (2012).

Chapter 4

Optical Properties of Multiferroic BiFeO$_3$ Films

Hiromi Shima, Hiroshi Naganuma and
Soichiro Okamura

Additional information is available at the end of the chapter

1. Introduction

1.1. Background

Lightwave communication systems are predominantly used for handling high-speed data traffic. Long-distance ground-based systems particularly depend on optical fibers. Several business and research facilities employ direct fiber connections, and fiber to the home (FTTH) technology is foreseeable in the near future. These developments are driven particulary by the high demand for bandwidth necessary for many computers contributing to internet traffic. Lightwave communication systems are one of the fastest growing industrial fields because of a few important inventions and extensive research and development by physicists and engineers. The key components of a long-distance lightwave communication system are semiconductor lasers, low-loss glass fibers, optical amplifiers, and photodetectors. Apart from these key elements, several additional functions are required to enable modulating, switching, and combining the optical signals. In addition, network traffic management and switching, routing, and distribution systems are essential. Therefore, we focused on the development of optical components such as modulators and switches.

Here, some of the novel optical devices for modulating or switching light signals are introduced. First, the operating principle of an electro-optic spatial light modulator (EOSLM) is described. In general, a spatial light modulator is an optical device that achieves spatial modulation of incident light. Figure 1(a) shows the basic structure of an optical switching cell in an EOSLM. [1, 2] An electro-optic thin film is fabricated on a large-scale integration (LSI) circuit together with top and bottom electrodes. The top electrode is made of transparent conductive material such as indium tin oxide (ITO). A dielectric multi-layer mirror (DMM) is deposited onto the ITO electrode, and a Fabry-Perot resonator is formed between the upper mirror and the bottom platinum electrode. In this case, the minimum reflectance becomes zero because the reflectance of the DMM and the bottom Pt layer is the same. If the refractive index can be controlled by applying an electric field, the cell can switch incident light at a specific wavelength as shown in Fig. 1(b).

Next, the operating principle of a Mach-Zehnder modulator (MZM) is described. Figure 2 shows a schematic diagram of an electrooptic-type MZM. [3-5] Incident light is split into two waveguides. The output amplitude depends on the phase difference at recombination. As shown in the top right part of the figure, in-phase recombination produced a "1" bit output while anti-phase recombination produces a "0" bit output owing to half-wave phase shifting caused by an applyied voltage. A half-wave phase shift can be caused by either the electro-optic effect or the thermo-optic effect. When using the thermo-optic effect, the optical switch structure includes a heater on one side of the waveguide instead of a top electrode. [6,7]

To realize these novel thin-film optical devices, many researchers have intensively studied materials that exhibit the electro-optic effect. Traditional electro-optic materials include (Pb,La)(Zr,Ti)O₃ (PLZT) and LiNbO₃. PLZT is the most promising candidate for such applications because it has a high transparency in its polycrystalline form. [7] It is well known that bulk PLZT with a 65:35 Zr:Ti ratio shows large electro-optic e • ect coe • cients: a Pockels coe • cient (linear electro-optic coe • cient) of 6.12×10^{-10} m/V and a Kerr coe • cient (quadratic electro-optic coe • cient) of 9.12×10^{-16} m²/V² with La contents of 8 and 9 at.%, respectively. [8] The next generation of new multiferroic materials, such as BiFeO₃ (BFO), which exhibits a giant remanent polarization of 100 μC/cm² in the thin-film form, [9,10] is now reaching maturity and has recently attracted considerable attention because of its potential applications in novel multifunctional devices. In recent years, the electric and magnetic properties of BFO films have been a topic of intense research, with regard to their magnetoelectric (ME) effect [11-13], while few reports on their optical properties have been published. [14-16] Therefore, it is necessary to know the basic optical properties of BFO films, such as the optical constant and thermooptic property, for the development of various optical applications. Understanding the optical potential in multiferroic materials leads us to additional noble material selections and thus degrees of freedom.

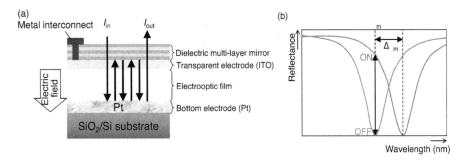

Figure 1. (a) Cross-sectional view of optical switching cell structure of the EOSLM, and 1(b) simulated optical switching properties of unit cells.

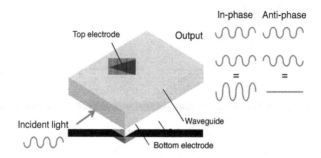

Figure 2. Schematic illustration of the electrooptic-type Mach-Zehnder modulator.

1.2. Objectives

To realize smaller and faster optical modulators or switches, it is necessary to embed electro-optic or thermo-optic materials in semiconductor integrated circuits. Accordingly, we have investigated the electro-optic and thermo-optic coefficients of materials in polycrystalline film form. We explain in detail the optical property evaluation method used in this study to exclude extrinsic effects, and then, we carefully determine the basic optical and thermo-optical properties of multiferroic $BiFeO_3$ polycrystalline films. Finally, we would like to understand the potential application of these optical properties to noble multifunctional devices using multiferroic $BiFeO_3$ polycrystalline films. This chapter describes the basic film preparation method and the basic method for evaluation of optical properties to increase understandings for beginner in this research field.

1.3. Outline of this chapter

This chapter is divided into 4 sections. Section 1 describes the background and objective of this study. In Section 2, we summarize the fundamentals of multiferroic $BiFeO_3$ used in this study, the method for fabricating multiferroic thin films, the basic optics principles related to this study, and the detailed method for evaluating optical properties. In Section 3, we discuss the thermooptic effect of multiferroic BFO films. The polycrystalline BFO films are fabricated on $Pt/Ti/SiO_2/Si$ substrates, respectively. First, their fundamental properties are evaluated. Next, the optical constants of these films are evaluated as a basic optical property. In Section 4, we summarize the conclusions of this chapter.

2. Experimental procedures

2.1. Fundamentals

Ferroelectricity is a property of certain materials that allows for spontaneous reversible electric polarization by applying an external electric field. The ability of a crystal to exhibit spontaneous polarization is related to its symmetry. Of the 32 point groups, which describe all crystalline systems, 11 are centrosymmetric and contain an inversion center. The remaining 21 point groups without an inversion center can exhibit piezoelectricity, except for the point group 432. Among the 21 groups without an inversion center, 10 polar groups possess a unique polar axis. Such asymmetric crystals can show unique electrical as well as optical characteristics, *e.g.*, the electro-optic effect, acousto-optic effect, photorefractive effect, nonlinear optical effect, etc. It is possible to control light dynamically using these effects to change the refractive index of materials.

In this chapter, the fundamentals of materials, fabrication, and evaluation are mentioned. In particular, the basic characteristics of BiFeO₃ potential ferroelectric films are mentioned together with their optical properties.

2.2. BiFeO₃ multiferroics

Perovskite-oxide has a structural formula of ABO_3, in which A is a large cation such as Bi^{3+}, Ba^{2+}, or Pb^{2+}, and B is a medium-sized cation such as Fe^{3+}, Ti^{4+}, or Zr^{4+}. These cations are located in cages formed by a network of oxygen anions, as shown in Fig. 3(a). Ferroelectric perovskites are a subgroup of the perovskite family. They are cubic at high temperatures and become polar non-cubic, *i.e.*, tetragonal, rhombohedral, *etc.*, below their Curie temperature. In the cubic phase, the cations are located at the center of an oxygen octahedron, while in the polar phases, they are shifted off center. The direction of the displacement of the cations in oxygen can be switched by applying an electric field, as shown in Fig. 3(b).

Multiferroic materials have more than one primary ferroic order parameter such as ferroelectricity, ferromagnetism, and ferroelasticity in the same phase. Multiferroic materials have attracted considerable attention, not only in terms of scientific interest but also because of their potential applications in novel functional devices. Bismuth ferrite (BiFeO₃, BFO) has long been known to be a multiferroic material that exhibits antiferromagnetism ($T_N \approx 643$ K) [17] and ferroelectricity ($T_C \approx 1103$ K) [18] when in bulk form. The structure and properties of thesingle-crystal form of BFO have been extensively studied. It has a rhombohedrally distorted perovskite structure with space group $R3c$ [19] at room temperature (RT), as shown in Fig. 4. A perovskite-type unit cell with a rhombohedral structure has a lattice parameter of $a = b = c = 0.3965$ nm and $\alpha = 89.3°$ at RT. [19,20]

The Fe magnetic moments are coupled ferromagnetically within the pseudocubic (111) planes and antiferromagnetically between the near planes; this is called the G-type antiferromagnetic order. If the magnetic moments are oriented perpendicular to the [111] direction, the symmetry also permits a canting of the antiferromagnetic sublattices resulting in a macroscopic magnetization called weak magnetism. [21,22]

According to a first-principles calculation, the spontaneous polarization of BFO changes depending on whether the crystal structure is rhombohedral or tetragonal. The tetragonal structure of the BFO (SG: $P4mm$) possesses P_s of around 150 μC/cm² along the [001] direction, and the rhombohedral structure (SG: $R3c$) possesses P_s of around 100 μC/cm² along the [111] direction without strain. [23,24] At the beginning of this research, in the case of bulk form, the

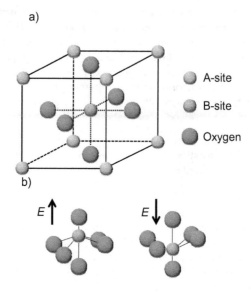

Figure 3. The crystal structure of typical perovskite-oxide: (a) The cubic phase, and (b) Schematic illustration of dipole switching with applying electric field $E > E_c$.

spontaneous polarization was 3.5 μC/cm^2 along the [001] direction, indicating a value of 6.1 μC/cm^2 along the [111] direction at 77 K. [25] Recently, it was reported that the bulk form of BFO showed P$_s$ 60 μC/cm^2 along the [111] direction 9er, in thin-film BFO, large remanent polarizations ranging from 100 to 150 μC/cm^2 have been reported. [9,10] The difference between the thin-film and bulk values, initially attributed to epitaxial strain, could also result from mechanic constraints in granular bulk ceramics and from leakage effects in crystals caused by defect chemistry or the existence of second phases.

2.3. Fabrication method

2.3.1. Preparation method using chemical solution deposition

Chemical solution deposition (CSD) [26] is the method for fabrication of thin films using a precursor solution; several types of metal-organic compounds such as metal alkoxide and metal carboxylate compounds can be used as the precursor solution. The fabrication of thin films by this approach involves four basic steps:

i.   Synthesis of the precursor solution;

ii.   Deposition by spin-casting or dip-coating, where the drying processes usually depends on the solvent;

iii.   Low-temperature heat treatment for drying, pyrolysis of organic species (typically 300–400°C), and formation of an amorphous film;

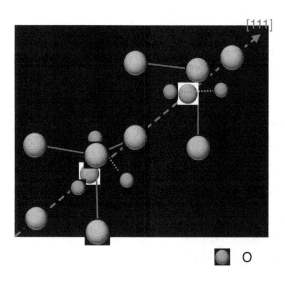

Figure 4. Schematic view of the *R3c* structure built up from two cubic perovskite BiFeO₃ unit cells. The cations are displaced along the [111] direction relative to the anions, and the oxygen octahedral rotate with alternating sense around the [111] axis.

iv.   Higher-temperature heat treatment for densification and crystallization of the coating into the desired oxide phase (600–1100°C).

For most solution deposition approaches, the final three steps are similar despite differences in the characteristics of the precursor solution, and for electronic devices, spin-casting has been used almost exclusively. Depending on the solution route employed, different deposition and thermal processing conditions may be used for controlling film densification and crystallization in order to prepare materials with optimized properties.

For the fabrication of perovskite thin films, the most frequently used CSD approaches may be grouped into three categories:

i.   Sol-gel processes that use 2-methoxyethanol as a reactant and solvent.

ii.   Chelate processes that use modifying ligands such as acetic acid.

iii.   Metal-organic decomposition (MOD) routes that use water-insensitive metal carboxylate compounds.

Other approaches that have also been used, although less extensively, include the nitrate method, citrate route, and Pechini process. In this study, a sol-gel solution and an enhanced-MOD (EMOD) solution (symmetric) were used because of their manageability.

2.3.2. Synthesis of the precursor solution for CSD

Processes based on 2-methoxyethanol are most appropriately considered sol-gel processes, and the key reactions leading to the formation of the precursor solutions are hydrolysis and condensation of the alkoxide reagents, in which metal-oxygen-metal (M-O-M) bonds are formed:

Hydrolysis:

$$M(OR)_nH_2O \rightarrow M(OR)_{n-1}(OH)ROH.$$

Condensation:

$$M(OR)_nM(OR)_{n-1}(OH) \rightarrow M_2O(OR)_{2n-2}ROH.$$

In some cases, an alcohol exchange reaction occurs in a practical synthesis process.

Alcohol exchange:

$$M(OR)_nxR'OH \rightarrow M(OR)_{n-x}(OR')_xxROH.$$

where OR is a reactive alkoxy group and OR' is the less reactive methoxyethoxy group.

In addition, to prepare a compound oxide material using two or more kinds of metal compounds in order to increase the homogeneity of the precursor solution, a double alkoxide with M-O-M'-O bonds may be synthesized for refluxing in an inactive gas atmosphere.

Synthesis of double alkoxide:

$$xM(OR)_mM'(OR')_n \rightarrow xMM'(OR)_m(OR')_n$$

In addition, if a metal alkoxide and carboxylate compounds are used, the synthesis is occasionally accompanied by an esterification reaction.

Esterification reaction:

$$M(OR)_nM'(O_2CR')_m \rightarrow (OR)_{n-1}MOM'RO_2CR'(O_2CR')_{m-1}RO_2CR'$$

For reproducible thin films, byproducts such as esters produced during the synthesis should be removed from the precursor solution through fraction distillation.

2.3.3. Preparation conditions of BFO polycrystalline films

The BFO films were formed through CSD. A precursor solution for CSD was prepared from bismuth acetate (99.99%, Aldrich), iron acetylacetonate (99.9%, Wako) in a solvent of 2-methoxyethanol (99.7%, Aldrich), and acetic acid (99.5%, Wako). The solution was heated to 80–100°C while stirring for 30 min to promote the dissolution of the precursors, followed by stirring at RT for 1 day before film deposition. The solution was synthesized on the basis of a stoichiometric composition. The concentration of the precursor solution was adjusted to be 0.05 M.

Figure 5 shows the process flow for the fabrication of the polycrystalline BFO film through CSD. A (111)Pt/Ti/SiO$_2$/(100)Si substrate was spin coated with the precursor solution at 3000 rpm for 30 s. The spin-coated film was dried at 150°C for 1 min and pyrolized at 400°C for 2 min in air. After the processes from spin coating to pyrolysis were repeated 10 times, the film was fired at 550°C for 5 min in air through rapid thermal annealing (RTA). This sequence was repeated 10 times. The film was polycrystalline with a random orientation and was approximately 650 nm thick.

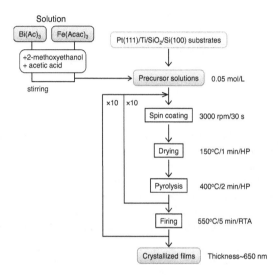

Figure 5. The process flow for fabrication of BFO films by CSD method.

2.4. Optical properties of ferroelectrics

2.4.1. Refractive index

The general definition of refractive index n is:

$$n = \frac{c_0}{v} \tag{1}$$

where c_0 is the speed of light in vacuum, and v, the speed of light in a material. The refractive index is related to the dielectric constant ε_r through the following equation:

$$\varepsilon_r = n^2 \tag{2}$$

This relationship is only valid when the interacting electric field has a frequency on the order of THz or higher and when the material is isotropic. The general behavior of condensed matter in an alternating electric field is that moving charges cause a frequency-dependent phase shift between the applied field and the electric displacement. Mathematically, this is expressed by writing the permittivity ε as a complex function:

$$\varepsilon(\varpi) = \varepsilon_1(\varpi) - i\varepsilon_2(\varpi) \tag{3}$$

The real part ε_1 characterizes the electric displacement, and the imaginary part ε_2 denotes the dielectric losses. The loss tangent is defined as

$$\tan\delta = \frac{\varepsilon_2}{\varepsilon_1} \tag{4}$$

Since light is an alternating electromagnetic wave with the electric and magnetic field vibration directions perpendicular to one another, the electric field induces an electric polarization in a dielectric crystal and the light itself is influenced by the crystal. The alternating frequency of light is so high (λ= 500 nm corresponds to a frequency of approximately 600 THz) that only electronic polarization can follow the electric field change. Therefore, the relative permittivity of an optically transparent crystal is small, typically less than 10. It is known that a dielectric material shows wavelength dispersion of its refractive index at optical frequencies.

2.4.2. Optical indicatrix and anisotropy of refractive index

In a microscopically anisotropic medium, the refractive index is generally different for different crystal directions. Ferroelectric materials, particularly in film form, can be both optically isotropic and optically anisotropic. Ferroelectric ceramics or polycrystalline films are an example of the former type; their isotropic behavior is due to the random orientation of their constituent grains. Ferroelectric single crystals or epitaxial films are an example of the latter type, and they can be divided into optically uniaxial and optically biaxial crystals. If a coordinate system is chosen to coincide with the three principal axes of a crystal, we have the following relations:

$$\varepsilon_x = n_x{}^2, \quad \varepsilon_y = n_y{}^2, \quad \varepsilon_z = n_z{}^2. \tag{5}$$

The optical anisotropy of a crystal is characterized by an optical indicatrix (or index ellipsoid) defined as

$$\frac{x^2}{n_x{}^2} + \frac{y^2}{n_y{}^2} + \frac{z^2}{n_z{}^2} = 1, \tag{6}$$

where n_x, n_y, and n_z are the principal refractive indexes, as shown in Fig. 2.6. The optical indicatrix is mainly used to find the two refractive indexes associated with the two independent plane waves that can propagate along an arbitrary direction k in a crystal. The optical indicatrix is used as follows: The intersection ellipse between the optical indicatrix and a plane through the origin point normal to the propagation direction k is found. The two axes of the intersection ellipse are equal in length to $2n_1$ and $2n_2$, where n_1 and n_2 are the two refractive indexes.

In the case of a biaxial crystal, there are two optical axes, and the refractive indexes are different in all three principal directions, $n_x \neq n_y \neq n_z$. In the common case of a uniaxial crystal, we have $n_x = n_y = n_o$ and $n_z = n_e$, where n_o and n_e are the ordinary and extraordinary refractive indexes, respectively. The refractive index along the optical axis corresponds to the extraordinary index n_e, and the refractive index perpendicular to the optical axis corresponds to the ordinary index n_o, as shown in Fig. 6. The existence of two rays with different refractive indexes is called birefringence. The birefringence Δn is usually defined as:

$$\Delta n = n_e - n_o \tag{7}$$

Since the value of n_e may be either higher or lower than n_o, birefringence may take on positive or negative values. If $\Delta n > 0$, the crystal is said to be optically positive, whereas if $\Delta n < 0$, it is said to be optically negative. For light propagating in a different direction from the principal axis in a uniaxial crystal, the situation becomes more complicated. A light wave with the wave vector κ, as shown in Fig. 6, has a constant ordinary index, whereas the extraordinary refractive index is dependent on the angle θ as:

$$\frac{1}{n_e(\theta)^2} = \frac{\cos^2\theta}{n_o^2} + \frac{\sin^2\theta}{n_e^2}. \tag{8}$$

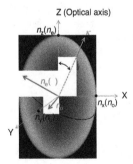

Figure 6. Schematic view of the *R3c* structure built up from two cubic perovskite BiFeO₃ unit cells. The cations are displaced along the [111] direction relative to the anions, and the oxygen octahedral rotate with alternating sense around the [111] axis.

2.4.3. Thermooptic effect

The thermo-optic effect refers to thermal modulation of the refractive index of a material. The refractive index of a material can be modulated as a function of its thermooptic coefficient α as

$$n(T) = n_0 - \alpha \cdot T, \tag{9}$$

where T is the temperature and $n(T)$ and n_0 are the refractive indexes at an arbitrary temperature and at 0°C, respectively. The thermo-optic coefficient relates the changes in the optical indicatrix ΔB_{ij} with the changes in temperature ΔT. Since temperature is a scalar, the thermo-optic effect is a symmetric second-rank tensor similar to the dielectric constant. The temperature dependence of the refractive index is generally small, except near phase transformations. The situation is analogous to low-frequency dielectrics. For silica and alumina, the permittivity is nearly independent of temperature, but ferroelectrics exhibit enormous changes near T_c. The refractive index of common oxides increases with density. Because of thermal expansion, density decreases with increasing temperature, thus decreasing the refractive index as well. Thermal expansion makes a small negative contribution to the temperature coefficient of refractive index dn/dT. This effect is often influenced by changes in the electronic band gap or by phase changes. These effects can be either positive or negative, depending on the nature of the energy levels or on the location of the phase transformation.

2.5. Evaluation method of optical properties

Spectroscopic ellipsometry

Ellipsometry determines the optical constants and thickness of materials in layered samples by fitting a parameterized model to the measured data for simultaneously analyzing data from multiple samples. Figure 7 shows a schematic illustration of ellipsometry. The linearly polarized incident light is reflected after interacting with the sample. The polarization of the light changes from linear to ellipsoidal from this interaction. We measure the polarization state using the ratio of the reflection coefficients for the light polarized parallel (p) and perpendicular (s) to the plane of incidence. This ratio, called the ellipsometric parameter, is defined as [27-30]

$$\rho = \frac{R_p}{R_s} = \tan(\psi)\exp(i\Delta), \tag{10}$$

where R_p and R_s are the Fresnel reflection coefficients of polarized light parallel and perpendicular to the incident plane, respectively. Here, tan (ψ) and Δ are the amplitude and phase of ρ, respectively. The Fresnel reflection coefficients are represented as follows:

For p-polarized light,

$$R_p = \frac{\tilde{n}^2 \cos\varphi - \sqrt{\tilde{n}^2 - \sin^2\varphi}}{\tilde{n}^2 \cos\varphi + \sqrt{\tilde{n}^2 - \sin^2\varphi}}, \tag{11}$$

and for s-polarized light,

$$R_s = \frac{\cos\varphi - \sqrt{\tilde{n}^2 - \sin^2\varphi}}{\cos\varphi + \sqrt{\tilde{n}^2 - \sin^2\varphi}}. \tag{12}$$

Here, $\tilde{n}$ is the complex refractive index, and Φ, the incident angle.

In this study, a Gaussian oscillator was used to model a dielectric function to represent film properties. Gaussian oscillators represent the normal distribution for the ε_2 spectrum; the ε_1 spectrum is determined by the ε_2 values because the Kramers-Kronig (KK) relation couples the real and imaginary parts of the complex dielectric constant. When a Gaussian oscillator is used as a dielectric function, complex dielectric constants (ε_1, ε_2) are calculated as [31,32]

$$\varepsilon_2 = A_n \exp\left[-\left(\frac{E - E_n}{B_n}\right)\right] + A_n \exp\left[-\left(\frac{E + E_n}{B_n}\right)\right], \tag{13}$$

$$\varepsilon_1 = 1 + \frac{2}{\pi} P \int_0^\infty \frac{\xi \varepsilon_2(\xi)}{\xi^2 - E^2} d\xi, \tag{14}$$

where A_n is the amplitude of the oscillator; E_n, the central energy of the oscillator; B_n, the broadening of the oscillator; and P, Cauchy's principal value. Optical constants (n, k) are equivalent to the complex dielectric constants (ε_1, ε_2). Therefore, optical constants (n, k) can be determined from the Gaussian oscillator parameters, which show the best fitting for experimental values of Φ and Δ. Model fitting was carried out by minimizing the mean square error (MSE) function defined as [27,29]

$$MSE = \frac{1}{2N - M} \sum_{i=1}^{N} \left[\left(\frac{\psi_i^{mod} - \psi_i^{exp}}{\sigma_{\psi_i}^{exp}}\right)^2 + \left(\frac{\Delta_i^{mod} - \Delta_i^{exp}}{\sigma_{\Delta_i}^{exp}}\right)^2 \right], \tag{15}$$

where N is the number of (ψ, Δ) pairs, M is the number of variable parameters in the model, σ is the standard deviation on the experimental points, and the superscripts "mod" and "exp" represent the calculated and experimental values, respectively.

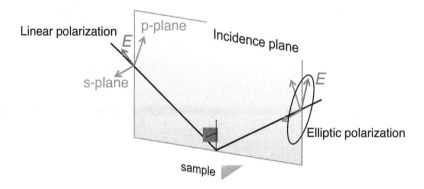

Figure 7. Schematic illustration of the ellipsometry.

3. Optical properties of multiferroic BiFeO$_3$ films

3.1. Fundamentals

Multiferroic materials, which simultaneously exhibit ferroelectricity and magnetic ordering, have attracted considerable attention, not only in terms of scientific interest but also because of their potential applications in novel functional devices. BFO is one of the few materials that exhibit dielectric and magnetic ordering at room temperature. [17,18] BFO also exhibits a large remanent polarization of 100 μC/cm^2 in thin film form. [9,10] Therefore, the electric and magnetic properties of BFO films have been the subject of intense research in recent years. [11-13] However, few reports on their optical properties [14,15] or their applications [16] have been published. It is important to know the exact optical properties of BFO films in order to develop various optical applications. To apply BFO films to optical devices, the electro-optic, magneto-optic, and thermo-optic effects of the films can be controlled by modulating their refractive indices. Recently, a Mach-Zehnder-type optical switch, which employs the thermo-optic effect, is a topic of immense interest in the photonics field. [33-35] In this section, we examine the optical constant and the temperature dependence of the refractive index of polycrystalline BFO films.

3.2. Experimental machines

- Crystal structure and orientation: X-ray diffractometer (PANalytical, X'pert PRO MPD)

- Morphology: field emission scanning electron microscope (JEOL, JIB-4500FE), transmission electron microscope (Hitachi, HF-2000), atomic force microscope (SII, SPI3800N)

- Electrical property: ferroelectric test system (Toyo, FCE)

- Optical property: spectroscopic ellipsometer (J. A. Woollam, M-2000) with a heating stage, as shown in Fig. 8(a) and 8(b).

a)

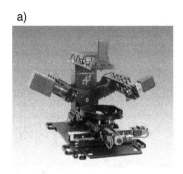

b)

Figure 8. (a) Spectroscopic ellipsometer (J. A. Woollam, M-2000) used in this study, and (b) the heating stage used for annealing sample from RT to 600°C.

3.3. Fundamental properties of polycrystalline BFO film

Polycrystalline BFO films were successfully formed on the Pt/Ti/SiO$_2$/Si substrates through CSD. Figure 9 shows the XRD θ-2θ pattern of the polycrystalline BFO film. The XRD analysis, it was confirmed that the BFO film was crystallized into a single perovskite phase with a random orientation.

Figure 10 shows the P-E hysteresis loop of the polycrystalline BFO film, measured with a single triangular pulse of 100 kHz at room temperature. From this figure, the ferroelectricity of the film can be confirmed. By a positive-up-negative-down (PUND) measurement technique [36] using a pulse train with an amplitude of 1.53 MV/cm and a width of 5 µs, a remanent polarization of 30 µC/cm^2, relative permittivity of 280, and leakage current density of 7.6 A/cm^2 were estimated at room temperature.

Figure 11 shows a) the surface morphology and the cross-sectional images of (b) the cleavage face and (c) the worked surface using a focused ion beam. From Fig. 11(a) and (b), it can be seen that the polycrystalline BFO film consists of small randomly grown grains, whereas the bottom Pt layer has a columnar grain growth. From Fig. 11(c), voids in the film were confirmed. Film thickness was estimated to be approximately 650 nm from cross-sectional images.

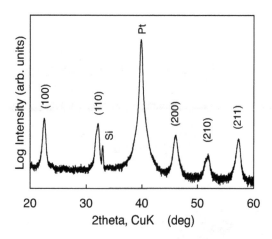

Figure 9. The XRD θ-2θ pattern of the polycrystalline BFO film.

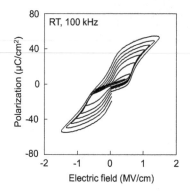

Figure 10. The P-E hysteresis loop of the polycrystalline BFO film measured with a single triangular pulse of 100 kHz at room temperature.

3.4. Optical constants of polycrystalline BFO film

Ellipsometric spectra in (Δ, ψ) were recorded at tincident angles of θ_i = 50°, 60°, and 70° in a spectral range of 245–1670 nm. Figure 12 shows the multilayer model used in this study. It was assumed that the model consisted of ambient (air), a surface layer, a bulk layer, and a substrate (Pt). The optical constants of the surface layer were represented by the Bruggeman effective medium approximation (EMA) [37] consisting of a 0.50 bulk film/0.50 void mixture. In the polycrystalline BFO film, 4 Gaussian oscillators were assumed to represent the film properties.

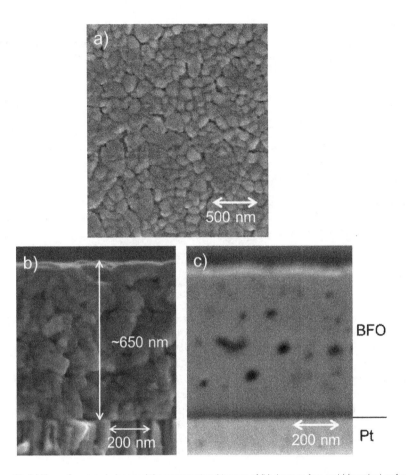

Figure 11. (a) The surface morphology and the cross-sectional images of (b) cleavage face and (c) worked surface by focused ion beam.

Furthermore, the density gradient along the film thickness, which affects the refractive index gradient, was also introduced by applying the EMA. Figure 13(a)–(d) shows the ψ and Δ spectra of the polycrystalline BFO film measured at various incident angles and fitting curves. The fitting parameters are summarized in Table 1. Figure 13(a) and (b) shows the results without the assumption of the refractive index gradient. In this case, the fitting curve did not represent the experimental result well and the MSE was relatively large. Figure 13(c) and (d) shows that when the refractive index gradient was considered, the MSE decreased from 106.6 to 66.6, and the fitting curves represented the experimental results well. This refractive index gradient seems to be caused by the distribution of voids in the film, as shown in Fig. 11(c). The total thickness of the polycrystalline BFO film was estimated to be approximately 650 nm. This value coincided with that observed in the cross-sectional SEM image shown in Fig. 11(a).

Figure 14 shows the optical constant of the polycrystalline BFO film in the wavelength range of 245 - 1670 nm calculated from the best-fitting results. In Fig. 14, the solid and broken lines represent the maximum and minimum values, respectively, of each optical constant in the graded layer. A large refractive index of 3.22 was estimated for the polycrystalline BFO film; this value is higher than that of rutile-type TiO$_2$ film, which has the highest refractive index among oxides reported thus far. At a wavelength of 600 nm, the refractive index of our BFO film was 3.22 and that of the TiO$_2$ film was reported to be approximately 2.6. [38] Figure 15 shows the depth profile of the refractive index and the extinction coefficient of the BFO film at a wavelength of 500 nm. This profile shows that the refractive index near the substrate is larger than that at the surface area. The reason for this gradient is not yet clear, although one possible explanation is the existence of voids in the film.

To determine the optical band gap, we plot $(\alpha E)^2$ vs. E for the polycrystalline BFO film, as shown in Fig. 16(a), where α and E are the absorption coefficient and photon energy, respectively. The absorption coefficient α is given by

$$\alpha = \frac{4\pi k}{\lambda},\tag{16}$$

where k is the extinction coefficient and λ, is the wavelength. A good linear fit above the band gap indicates that the BFO film has a direct gap. [15] The linear extrapolation of $(\alpha E)^2$ to 0 induces band gaps of 2.79 eV for the polycrystalline BFO film. This means that the absorption edge of the BFO film is 445 nm. This value for the polycrystalline BFO film was similar to that in recent reports. [15,39,40] Next, we plot $(\alpha E)^{1/2}$ vs. E in Fig. 16(b) for the BFO film. The $(\alpha E)^{1/2}$ vs. E plot did not show 2 clear slopes as was expected for an indirect gap material. [41] Finally, the polycrystalline BFO film was found to show a sufficiently low light loss at a wavelength greater than 600 nm so that its large reflective index at visible wavelengths is useful for electric-optic devices such as an electro-optic spatial light modulator.

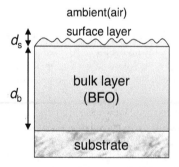

Figure 12. A multilayer model used in this study.

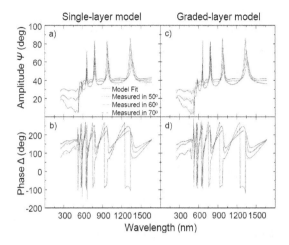

Figure 13. Ψ and Δ spectra of the polycrystalline BFO film measured at various incident angles and fitting curves obtained using the model (a) - (b) without and (c) - (d) with refractive index gradient.

$$\varepsilon_1 = 1 + \frac{2}{\pi} P \int_0^\infty \frac{\xi \varepsilon_2(\xi)}{\xi^2 - E^2} d\xi, \quad \varepsilon_2 = A_n \exp\left[-\left(\frac{E - E_n}{B_n}\right)\right] + A_n \exp\left[-\left(\frac{E + E_n}{B_n}\right)\right]$$

Fitting parameters		Sample-Model	
		BFO-single	BFO-graded
Thickness (nm)	d_s	3.566±0.635	4.597±0.224
	d_b	667.423±3.31	644.839±1.92
Amplitude A_n	A_{n1}	6.771	9.834
	A_{n2}	8.981	10.973
	A_{n3}	3.638	5.008
	A_{n4}	0.735	0.716
Central energy E_n (eV)	E_{n1}	4.175	4.119
	E_{n2}	7.460	7.348
	E_{n3}	3.024	3.052
	E_{n4}	2.491	2.476
Broadening B_n (eV)	B_{n1}	1.201	1.205
	B_{n2}	4.254	4.254
	B_{n3}	0.527	0.527
	B_{n4}	0.215	0.215
MSE		106.6	66.66

Table 1. The fitting parameters in single layer model and graded layer model of the polycrystalline BFO films. Gaussian oscillator defined as following equation.

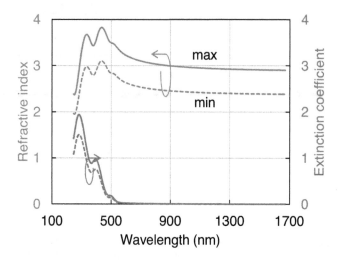

Figure 14. Refractive index and extinction coefficient of the polycrystalline BFO film; the solid and broken lines show maximum and minimum values, respectively.

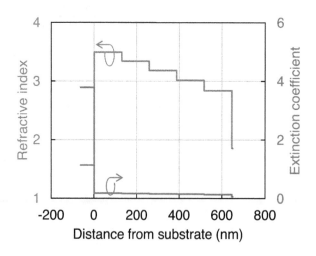

Figure 15. Depth profile of refractive index and extinction coefficient of the BFO film at a wavelength of 500 nm.

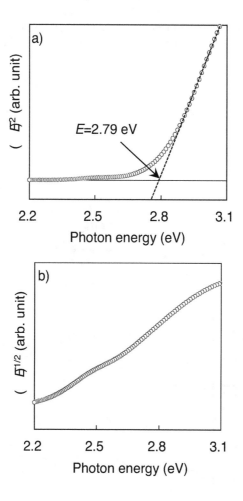

Figure 16. Plots of (a) $(\alpha E)^2$ and (b) $(\alpha E)^{1/2}$ vs. photon energy E for the polycrystalline BFO film. The linear extrapolation of $(\alpha E)^2$ to 0 gives band gaps of 2.79 eV.

3.5. Thermooptic property of polycrystalline BFO film

3.5.1. Temperature dependence of lattice space

Figure 17 shows the lattice spacing of Si (400), BFO (100), and Pt (111) as a function of temperature, estimated from XRD patterns. All the d-spaces monotonically increased with increasing temperature because of thermal expansion. Thermal expansion coefficients, estimated from Fig. 17, are shown in Table 2, which includes reference data [42-44] for comparison. From this table, we can see that our experimental values are larger than those in reference data, except for the Si substrate because of the effect of in-plane compressive stress.

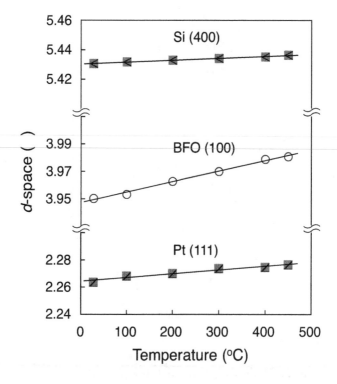

Figure 17. The lattice spacing of Si (400), BFO (012), and Pt (111) as a function of temperature estimated from XRD patterns.

Material	Thermal expansion coefficient	
	Experimental data ($\times 10^{-6}$ K^{-1})	Reference data ($\times 10^{-6}$ K^{-1})
BFO	19	10-14
Pt	12	9
Si	2.4	2

Table 2. The thermal expansion coefficients of the BFO, Pt, and Si.

3.5.2. Temperature dependence of refractive index

Ellipsometric spectra in (Δ, Ψ) were recorded at incident angles of $\theta_i = 70°$ in a spectral range of 245–1670 nm. The sample was loaded onto the customized heating stage and heated from RT to 530°C. In these measurements, one Gaussian oscillator was assumed to represent film properties, and the central energy E_n of the oscillator was fixed at 5 eV for each temperature. Figure 18(a) – b) shows the wavelength dispersion of the refractive index and the extinction coefficient measured at 50 and 530°C, respectively. From this figure, we observe that the refractive index decreases with increasing temperature for all wavelengths. The refractive index variation with increasing temperature is large for shorter wavelengths. Based on these results, we investigated the origin of the temperature dependence of refractive index dispersion, as shown in Fig. 18. In the short-wavelength region, the lattice vibration becomes more intense with increasing temperature; therefore, in a high-temperature region, the amplitude of the oscillator decreased, and the broadening of the oscillator expanded. At the same time, density is decreased with increasing temperature owing to thermal expansion; therefore, the refractive index decreased in all wavelength regions. This combination seems to be responsible for the refractive index curve shown in Fig. 18(a).

Figure 19 shows the temperature dependence of the refractive index in a long- wavelength region. It is found that refractive index decreases with increasing temperature at each wavelength, although there was some variability. The thermo-optic coefficient was estimated from the slopes of linear approximation. A thermo-optic coefficient of 0.8×10^{-4} K^{-1} was obtained at a wavelength of 1550 nm for the BFO film. The thermo-optic coefficients of polymers and glasses, which are known to be typical thermo-optic materials, have been reported to be approximately 2×10^{-4} K^{-1} and 0.1×10^{-4} K^{-1}, respectively. [45,46] It is found that the BFO film shows a large refractive index and a thermo-optic coefficient comparable to that of these traditional materials. Finally, it can be concluded that the BFO also has potential for use in an electro-optic-type Mach-Zehnder modulator.

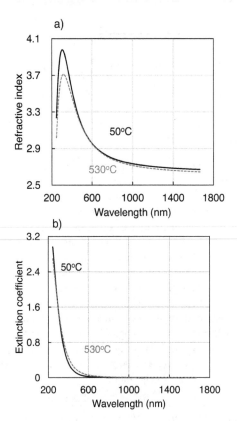

Figure 18. The wavelength dispersion of a) refractive index and b) extinction coefficient measured at 50 and 600ºC, respectively.

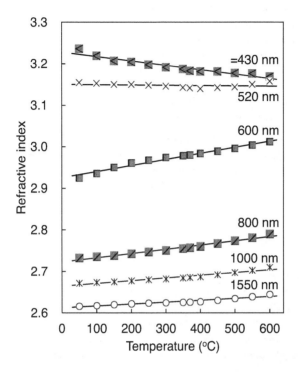

Figure 19. The refractive indexes at various wavelengths as a function of temperature.

4. Conclusions

The optical and thermo-optic properties of BiFeO$_3$ (BFO) films were studied. Polycrystalline BFO films were formed on Pt/Ti/SiO$_2$/Si substrates, and their basic optical and thermo-optic properties were systematically evaluated. The new findings are summarized as follows:

i.   The polycrystalline BFO films were evaluated using a spectroscopic ellipsometer. Gaussian oscillators were assumed as a dielectric function to represent film properties, and the graded model was assumed by introducing a refractive index gradient. As a result, large refractive indexes of 3.22 and 2.91 were estimated for the polycrystalline BFO film at wavelengths of 600 and 1550 nm, respectively, these refractive indexes are higher than that of the rutile-type TiO$_2$ film, which is known to have a high refractive index. The optical band gap of the BFO film at RT was estimated as a direct transition to be 2.79 eV, which corresponds to the absorption edge of 445 nm. It was found that the BFO film shows sufficiently low light loss at wavelengths larger than 600 nm.

ii.   The thermo-optic properties of the BFO films were evaluated using a spectroscopic ellipsometer with a heating stage. The refractive index of the polycrystalline film decreased with increasing temperature. We considered that this change in refractive index is caused by the balance between the increase in refractive index due to the enhancement of the oscillator dispersion and the decrease in refractive index due to the decreased density of the film. In addition, we obtained thermo-optic coefficient of $0.8 \times 10^{-4} K^{-1}$ at a wavelength of 1550 nm for the BFO film, which is larger than those of typical thermo-optic polymers ($2 \times 10^{-4} K^{-1}$) and glasses ($0.1 \times 10^{-4} K^{-1}$).

These results suggested that the BFO films have a high potential for application as an optical material with a high refractive index, and that the effectual refractive index change can be controlled by the balance of two factors, the activation of the oscillator and the thermal expansion coefficient, even in the same material.

Acknowledgements

The author thanks Assistant Prof. Takashi Nakajima, Tokyo University of Science, Prof. Akiharu Morimoto, Asso. Prof. Takeshi Kawae, Kanazawa University, Prof. Takashi Yamamoto, Ass. Prof. Ken Nishida, National Defense Academy, Dr. Takashi Iijima, AIST, Dr. Koichi Tsutsumi, Dr. Masahiro Matsuda, Dr. Michio Suzuki, J. A. Woollam Japan, and Dr. Toshiyasu Tadokoro Techno-Synergy, Inc. This study was partially supported by a Grant-in-Aid for JSPS Research Fellows (No. 217990) from the Japan Society for the Promotion of Science.

Author details

Hiromi Shima[1*], Hiroshi Naganuma[2*] and Soichiro Okamura[1]

*Address all correspondence to: shima@nda.ac.jp, naganuma@mlab.apph.tohoku.ac.jp

1 Department of Applied Physics, Tokyo University of Science, 1-3, kagurazaka, Shinjuku-ku, Tokyo, Japan

2 Department of Applied Physics, Graduate school of Engineering, Tohoku University, Japan

References

[1] Fujimori, Y.; Fujii, T.; Suzuki, T.; Kimura, H.; Fuchikami, T.; Nakamura, T.; and Takasu, H. (2005) *IEDM Tech. Dig.*, 2005 p.p. 935.

[2] Bitou, Y.; and Minemoto, T. (1998) High-contrast spatial light modulator by use of the electroabsorption and the electro-optic effects in a GaAs single crystal, *Appl. Opt.*, Vol. 37, August 1998, pp. 4347-4356.

[3] Thapliya, R.; Okano, Y.; and Nakamura, S. (2003) Electrooptic Characteristics of Thin-Film PLZT Waveguide Using Ridge-Type Mach–Zehnder Modulator, *J. Lightwave Tech.*, Vol. 21, January 2003 pp. 1820-1827.

[4] Jacobsen, R. S.; Andersen, K. N.; Borel, P. I.; Pedersen, J. F.; Frandsen, L. H.; Hansen, O.; Kristensen, M.; Lavrinenko, A. V.; Moulin, G.; Ou, H.; Peucheret, C.; Zsigri, B.; and Bjarklev, A. (2006) Strained silicon as a new electro-optic material, *Nature*, Vol. 441, May 2006 pp. 199-202.

[5] Shimizu, T.; Nakada, M.; Tsuda, H.; Miyazaki, H.; Akedo, J.; and Ohashi, K. (2009) Gigahertz-rate optical modulation on Mach-Zehnder PLZT electro-optic modulators formed on silicon substrates by aerosol deposition, *IEICE Electro. Exp.*, Vol. 6, December 2009 pp. 1669-1675.

[6] Xie, N.; Hashimoto.; T and Utaka, K. (2009) Very Low-Power, Polarization-Independent, and High-Speed Polymer Thermooptic Switch *IEEE Photonics Tech. Lett.*, Vol. 21, December 2009 pp. 1861-1863.

[7] Reed, G. T.; Mashanovich, G.; Gardes, F. Y.; and Thomson, D. J.; (2010) Silicon optical modulators, *Nature Photonics*, Vol. 4, July 2010 pp. 518-526.

[8] Haertling, G. H,; and Land, C. E.; (1971) Hot-Pressed (Pb, La)(Zr,Ti)O₃, Ferroelectric Ceramics for Electrooptic Applications*J. Am. Ceram. Soc.*, Vol. 54, October 1971, pp. 1-11.

[9] Wang, J.; Neaton, J. B.; Zheng, H.; Nagarajan, V.; S. Ogale, B.; Liu, B.; Viehland, D.; Vaithyanathan, V.; Schlom, D. G.; Waghmare, U. V.; Spaldin, N. A.; Rabe, K. M.; Wuttig, M. & Ramesh, R. (2003) Epitaxial BiFeO₃ Multiferroic Thin Film Heterostructures, *Science* Vol. 299, February 2003, pp. 1719-1722.

[10] Yun, K. Y.; Rincinschi, D.; Kanashima, T.; Noda, M.; and Okuyama, M.; (2004) Giant Ferroelectric Polarization Beyond 150 µC/cm² in BiFeO₃ Thin Film*Jpn. J. Appl. Phys.* Vol. 43, 2004 pp. L647-L648.

[11] Bibes, M.; and Barthelemy, A.; (2008) Towards a magnetoelectric memory*Nature materials*, Vol. 7, 2008 pp. 425-426.

[12] Ramesh, R.; and Spaldin, N. A. (2007) Multiferroics: progress and prospects in thin films *Nature materials*, Vol. 6, 2007 pp. 21-29.

[13] Chu, Y. H.; Martin, L. W.; Holcomb, M. B.; Gajek, M.; Han, S. –J.; He, Q.; Balke, N.; Yang, C. –H.; Lee, D.; Hu, W.; Zhan, Q.; Yang, P. –L.; Rodriguez, A. F.; Scholl, A.; Wang, S. X.; and Ramesh, R.; (2008) Electric-field control of local ferromagnetism using a magnetoelectric multiferroic*Nature materials*, Vol. 7, 2008, pp. 478-482.

[14] Iakovlev, S.; Solterbeck, C. –H.; Kuhnke, M.; and Es-Souni, M.; (2005) Multiferroic Bi-FeO$_3$ thin films processed via chemical solution deposition: Structural and electrical characterization *J. Appl. Phys.*, Vol. 97, 2005 pp. 094901-1-094901-6.

[15] Kumar, A.; Rai, R. C.; Podraza, N. J.; Denev, S.; Ramiez, M.; Chu, Y. –H.; Martin, L. W.; Ihlefeld, J.; Heeg, T.; Schubert, J.; Schlom, D. G.; Orenstein, J.; Ramesh, R.; Collins, R. W.; Musfeldt, J. L.; and Gopalan, V.; (2008) Linear and nonlinear optical properties of BiFeO$_3$ *Appl. Phys. Lett.*, Vol. 92, 2008, pp. 121915-1-121915-3.

[16] Choi, T.; Lee, S.; Choi, Y. J.; Kiryukhin, V.; and Cheong, S. –W. (2009) Switchable Ferroelectric Diode and Photovoltaic Effect in BiFeO$_3$ *Science* Vol. 324, 2009, pp. 63-66.

[17] Kiselev, S. V.; Ozerov, R. P.; and Zhdanov, G. S.; (1963) Sov. Phys., Vol. 7 1963, pp. 742.

[18] Venevtsev, Yu. N.; Zhadanow, G.; and Solov'ev, S.; (1960) Sov. Phys. Crystallogr., Vol. 4 1960, pp. 538.

[19] Kubel, F.; and Schmid, H.; (1990) Structure of a Ferroelectric and Ferroelastic Monodomain Crystal of the Perovskite BiFeO$_3$ *Acta Crystallogr. B*, Vol. 46 1990, pp. 698-702.

[20] Filip'ev, V. S.; Smol'yaninov, I. P.; Fesenko. E. G.; and Belyaev, I. I.; Kristallografiya, Vol. 5 1960 pp. 958.

[21] Dzyaloshinskii, I. E.; S(1957) Sov. Phys. JETP, Vol. 5 1957 pp. 1259.

[22] Moriya, T. (1960) Anisotropic superexchange interaction and weak ferromagnetism *Phys. Rev.*, Vol. 120, 1960 pp. 91-98.

[23] Neaton, J. B.; Ederer, C.; Waghmare, U. V.; Spaldin, N. A.; and Rabe, K. M.; (2005) First-principles study of spontaneous polarization in multiferroic BiFeO$_3$ *Rhys. Rev. B*, Vol. 71 2005 pp. 014113-1-014113-8.

[24] Ederer, C.; and Spaldin N. A. (2005) Effect of Epitaxial Strain on the Spontaneous Polarization of Thin Film Ferroelectrics, *Physical Review Letters.*, Vol. 95, December 2005, pp. 257601-1-4.

[25] Teague, J. R.; Gerson, R.; and James, W. J.; (1970) Dielectric Hysteresis in single crystal BiFeO$_3$ *Solid State Commun.*, Vol. 8 1970 pp. 1073-1074.

[26] Schwartz, R. W. (1997) Chemical Solution Deposition of Perovskite Thin Films *Chem. Mater.*, Vol. 9 1997 pp. 2325-2340.

[27] Aspnes, D. E. (1974) Optimizing precision of rotating-analyzer ellipsometers *J. Opt. Soc. Am.*, Vol. 64 1974 pp. 639-646.

[28] Jayatissa, A. H.; Suzuki, M.; Nakanishi, Y.; and Hatanaka, Y. (1995) Microcrystalline structure of poly-Si films prepared by cathode-type r.f. glow discharge *Thin Solid Films*, Vol. 256 1995 pp. 234-239.

[29] Yao, H.; Snyder, P. G.; and Woollam, J. A.; (1991) Temperature dependence of optical properties of GaAs *J. Appl. Phys.*, Vol. 70 1991 pp. 3261-3267.

[30] Yao, H.; Woollam, J. A.; and Alterovitz, S. A.; (1993) Spectroscopic ellipsometry studies of I-IF treated Si (100) surfaces *Appl. Phys. Lett.*, Vol. 62 1993 pp.3324-3326.

[31] Brendel, R.; and Bormann, D.; (1992) An infrared dielectric function model for amorphous solids *J. Appl. Phys.*, Vol. 71 1992 pp. 1-6.

[32] Kim, C. C.; Garland, J. W.; Abad, H.; and Raccah, P. M.; (1992) Modeling the optical dielectric function of semiconductors: extension of the critical0point parabolic-band approximation *Phys. Rev. B*, Vol. 45 1992 pp. 11749-11767.

[33] Khan, M. H.; Shen, H.; Xuan, Y.; Zhao, L.; Xiao, S.; Leaird, D. E.; Weiner, A. M.; and Qi, M. (2010) Ultrabroad-bandwidth arbitrary radiofrequency waveform generation with a silicon photonic chip-based spectral shaper *Nature Photonics*, Vol. 4 2010 pp. 117-122.

[34] Yamada, H.; Chu, T.; Ishida, S.; and Arakawa, Y. (2006) Si Photonic Wire Waveguide Devices *IEEE Journal of selected topics in quantum electronics*, Vol. 12 2006 pp. 1371-1378.

[35] Wang, X.; Xu, L.; Li, D.; Liu, L.; and Wang, W.; (2003) Thermo-optic properties of sol-gel-fabricated organic–inorganic hybrid waveguides *J. Appl. Phys.*, Vol. 94 2003 pp. 4228-4230.

[36] Naganuma. H.; Inoue, Y.; and Okamura, S.; (2008) Evaluation of Electrical Properties of Leaky BiFeO₃ Films in High Electric Field Region by High-Speed Positive-Up–Negative-Down Measurement *Appl. Phys. Exp.*, Vol. 1 2008 pp. 061601-1-061601-3.

[37] Landauer, R.; in Electrical Transport and Optical Properties of Inhomogeneous Media, ed. J. C. Garland and D. B. Tanner (AIP, New York, 1979) p. 1.

[38] Ting, C. –C.; Chen, S. –Y.; and Liu, D. –M.; (2000) Structural evolution and optical properties of TiO₂ thin films prepared by thermal oxidation of sputtered Ti films *J. Appl. Phys.*, Vol. 88 2000 pp. 4628-4633.

[39] Ihlefeld, J. F.; Podraza, N. J.; Liu, Z. K.; Rai, R. C.; Xu, X.; Heeg, T.; Chen, Y. B.; Li, L.; Collins, R. W.; Musfeldt, J. L.; Pan, X. Q.; Schubert, J.; Ramesh, R.; and Schlom, D. G. (2008) Optical band gap of BiFeO₃ grown by molecular-beam epitaxy *Appl. Phys. Lett.*, Vol. 92 2008 pp. 908-1-908-3.

[40] Hauser, A. J.; Zhang, J.; Mier, L.; Ricciardo, R. A.; Woodward, P. M.; Brillson, L. J.; and Yang, F. Y.; (2008) Characterization of electronic structure and defect states of thin epitaxial BiFeO₃ films by UV-visible absorption and cathodoluminescence spectroscopies *Appl. Phys. Lett.*, Vol. 92 2008 pp. 222901-1-222901-3.

[41] Pankove, J. I.; Optical Process in Semiconductors (Prentice-Hall, Englewood Cliffs, NJ, 1971) p. 34.

[42] Amirov, A. A.; Batdalov, A. B.; Kallaev, S. N.; Omarov, Z. M.; Verbenko, I. A.; Razu-movskaya, O. N.; Reznichenko, L. A.; and Shilkina, L. A.; (2009) Specific Features of the Thermal, Magnetic, and Dielectric Properties of Multiferroics $BiFeO_3$ and $Bi_{0.95}La_{0.05}FeO_3$ *Phys. Solid State*, Vol. 51 2009 pp. 1189-1192.

[43] Watanabe, H.; Yamada, N.; and Okaji, M. (2004) Linear Thermal Expansion Coefficient of Silicon from 293 to 1000 K *Inter. J. Thermophys.*, Vol. 25 2004 pp. 221-236.

[44] Okada, Y.; and Tokumaru, Y.; (1984) Precise determination of lattice parameter and thermal expansion coefficient of silicon between 300 and 1500 K *J. Appl. Phys.*, Vol. 56 1984 pp. 314-320.

[45] Zhang, Z.; Zhao, P.; Lin, P.; and Sun, F.; (2006) Thermo-optic coefficients of polymers for optical waveguide applications *Polymer*, Vol. 47 2006 pp. 4893-4896.

[46] Jewell, J. M.; Askins, C.; and Aggarwal, D. (1991) Interferometric method for concurrent measurement of thermo-optic and thermal expansion coefficients *Appl. Opt.*, Vol. 30 1991 pp. 3656-3660.

Chapter 5

Physicochemical Analysis and Synthesis of Nonstoichiometric Solids

V. P. Zlomanov, A.M. Khoviv and A.Ju. Zavrazhnov

Additional information is available at the end of the chapter

1. Introduction

Technological progress is intimately associated with creation of new materials, such as composites, piezoelectrics, ferroelectrics, semiconductors, superconductors and nanomaterials with preset functional properties. For the synthesis of these materials, it is necessary to study the chemical processes that lead to the composition, structure, and accordingly desired properties. Investigation of the interrelation between the composition, structure and properties of matter and determination of synthetic conditions for obtaining substances with preset composition and structure are the major problems of physicochemical analysis. Although significant progress has been made in understanding several challenges remain for further advancements. These challenges and new approaches include some definitions – stoichiometry, nonstoichiometry, deviation from stoichiometry, substance, phase, component as well as the use of phase diagrams in selecting conditions for the synthesis of nonstoichiometric solids. Since nonstoichiometry is associated with defects, attention is also paid to their classification and formation. Synthesys of solid involves control over phase transformations. For this reason some features of the $P-T-x$ phase diagrams are discussed. The following inportant features of $P-T-x$ phase diagrams will be considered: highest maximal melting point $T_{m,AB}^{max}$ of the solid compound S_{AB}, noncoincidence of the solid, liquid and vapor compositions ($x^L \neq x^S \neq x^V$) at this temperature, factors determining the value of the homogeneity range, some features of the terms congruent and incongruent phases and phase processes. Criteria for evaluating the homogeneity of nonstoichiometric solids are also considered.

The term *Physicochemical analysis* was introduced in [1] and defined as the field of chemistry dealing with the interrelation between the composition and properties of matter [1]. The foundations of physicochemical analysis were in [1-6].

It has been realized that it is necessary to study the thermodynamic properties of solids and the phase diagrams of the systems in which these solids occur. Because the properties of solids depend significantly on their composition, great attention has been focused on the physico-chemical analysis and foundations of the directed synthesis.

2. The essence of physicochemical analysis — Some definitions

Investigation of the interrelation between the composition, structure, and properties as well as determination of synthetic conditions for obtaining solids with preset composition and structure are the the basic problems of physicochemical analysis.

The subject of chemistry is the conversions of substances. What is a substance, and what is its conversion? A substance is a multitude of interacting particles possessing certain characteristics: composition, particle size, structure, and the nature of chemical bonding. It is these characteristics that determine the properties of the substance (Figure 1).

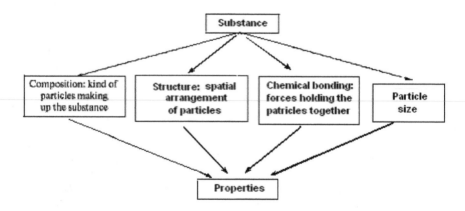

Figure 1. Interrelation between the basic characteristics and properties of a substance.

Composition is defined as the kinds of particles constituting the substance. For example, a sodium chloride crystal is built of sodium and chloride ions occupying cationic sites (Na_{Na}^{+}) and anionic sites (Cl_{Na}^{+}), respectively. The constituent particles can be not only atoms or ions, but also molecules (e.g., I_2 molecules in iodine crystals and water molecules in ice), coordination polyhedra (e.g., SO_4^{2-} tetrahedra in potassium sulfate), and other entities.

Structure is some ordered arrangement of the above particles in space.

The properties of a crystal, such as the lattice energy and electrical, optical, and chemical properties, are determined by the composition and structure of the crystal. Different spatial arrangements of the same particles, such as carbon atoms in diamond and graphite, are characterized by different lattice energies and, accordingly, different properties, including melting and boiling points and hardness.

The chemical bond is understood as the forces binding the particles together. These forces arise from the the Coulomb interaction of electrons and nuclei. Depending on the electron distribution among nuclei, there can be ionic, covalent, and metallic bonds.

The particle interaction energy depends on the particle size. At the nanometer level (1–100 nm), it changes markedly, making possible the formation of new physical and chemical properties of the substance. This is explicable in terms of surface physics and chemistry (dependence of the surface energy on the particle size).

A conversion of a substance is a change in one or several characteristics of the substance (Figure1). This process is accompanied by a change in energy (dU) in the form of heat transfer (δQ) or work execution (δA):

$$dU = \delta Q - \delta A. \tag{1}$$

The main purpose of directed synthesis is to obtain substances with the preset composition, structure, and, hence, properties.

The direct synthesis of solids includes control of phase transitions. Therefore, thermodynamic data characterizing phases and their transitions are necessary to estimate the optimum synthetic conditions.

Now let us consider some specific features of the concepts of a phase and a component.

2.1. Phase and component

A substance is made up of particles or their interacting sets with a certain structure and chemical bonding. Energy is an equivalent or measure of these interactions. In thermodynamics, the state of a system is defined using a set of variables, or coordinates, such as pressure P, temperature T, and volume V. For a thermodynamic state of a system characterized by a set of coordinates (intensive thermodynamic properties), Gibbs suggested the term *phase of matter* [7, 8]. This definition emphasizes that the phase of matter is size- and shape-independent. Later, the word combination *phase of matter* gave way to the term *"phase."*

The equation of state of a phase in terms of pressure (P), temperature (T), and composition (x) is written as (2)

$$dG = -SdT + VdP + \sum_{i=1}^{k}\left(\frac{\partial G}{\partial x_i}\right)_{P,T,x_j} dx_i, \tag{2}$$

where G, S, and V are the molar Gibbs energy, molar entropy, and molar volume of the phase, espectively; x_i is the mole fraction of the i-th component k; x_j means the constancy of the mole fractions of all components but not the i-th component. The phases can be individual solid, liquid, gaseous compounds, or solid solutions, including nonstoichiometric compounds, whose thermodynamic properties are described by equations of state like Eq. (2) and are continuous functions of P, T, and x.

Note the following specific features of the concept of a phase. Firstly, existing phases should be distinguished from coexisting phases. The properties (Gibbs energy G, enthalpy, etc.) of an existing phase, such as a solution containing different concentrations of the same substance, are continuous functions of composition. By contrast, the properties of coexisting phases (i.e., phases that are in equilibrium) are equal in all parts of the system. The second specific feature is associated with the definition of a phase as the homogeneous part of a heterogeneous system. This definition is inexact because it does not specify whether homogeneous parts of a heterogeneous system are one phase or different phases. Thirdly, the concept of a phase is broader than the concept of a physical state (gas, solid or liquid). Finally, the fourth specific feature of the concept of a phase is associated with the issue of the minimum set of particles describable in terms of one equation of state like Eq. (2). From the standpoint of kinetic molecular theory, this set should be sufficiently large to allow regular particle energy distribution. Estimates demonstrate that the number of particles must be at least several tens. The properties of such a phase will be affected by surface tension [9].

The components of a system are the types of particles constituting this system. They are called constituents, and their number is designated n. If the concentrations of the n constituents are related by m independent equations, then $k = n - m$ is the number of independent constituents, or, simply, the number of components. For example, the system $NH_3(g) + HCl(g) = NH_4Cl(s)$, which consists of three substances ($n = 3$), whose concentrations are related by one equation ($m = 1$), is two-component since $k = n - m = 3 - 1 = 2$.

The components are subject to the following constraints:

1.   their concentrations must be independent of one another;

2.   they should completely describe the concentration dependence of the properties of the system;

3.   their number should satisfy the electroneutrality principle.

2.2. Stoichiometry, nonstoichiometry, deviation from stoichiometry

The properties of a substance depend on its composition. The great focus of materials sience are the concepts of stoichiometry, nonstoichiometry, and deviation from stoichiometry.

The proportions in which substances react are governed by stoichiometric laws (stoichiometry). These laws, which characterize the composition of chemical compounds, were discovered by systematizing experimental data. The fundamental laws of stoichiometry include the law of constant composition and the law of multiple proportions.

The law of constant composition, established in the 19th century by the French chemist Joseph Louis Proust, states that the chemical composition of a compound is independent of the way in which this compound was obtained. The law of multiple proportions, formulated by the English chemist John Dalton in 1807, states that, when two elements combine with each other to form more than one compound, the mass fractions of the elements in these compounds are in a ratio of prime numbers. Both laws follow from atomistic theory and suggest that the saturation of the chemical bonds is necessary for the formation of a molecule from atoms. Any change in the number of atoms or their nature or arrangement indeed means the formation of a new molecule with new properties.

Are the law of constant composition and the law of multiple proportions always obeyed? They are valid only for the substances constituted by molecules. In fact, the composition of a substance can vary significantly, depending on the preparation conditions. It was long believed that only those chemical substances exist whose composition obeys the law of multiple proportions. They are stoichiometric and are called daltonides in honor of John Dalton. However, as methods of investigation were making progress, it turned out that the properties of solid inorganic substances, such as vapor pressure, electric conductivity, and diffusion coefficients, are composition dependence. In some composition range, the structure, i.e. the arrangement of the components in space remains invariable, while the component concentrations vary continuously. This range is called the homogeneity range or the nonstoichiometry range. These substances are referred to as nonstoichiometric or variable-composition compounds. Earlier, they were called berthollides in honor of Claude Louis Berthollet, Proust's compatriot. A nonstoichiometric compound can be treated as a solid solution of its components, such as cadmium and tellurium in the compound CdTe.

The homogeneity range is characterized by the corresponding deviation from stoichiometry. The stoichiometric composition of a solid compound, e.g., A_nB_m, where n and m are prime numbers, is the composition that obeys the law of multiple proportions. The deviation from stoichiometry or, briefly, nonstoichiometry (Δ) is defined as the difference between the ratio of the number of nonmetal atoms B to the number of metal atoms A in the real $A_nB_{m+\delta}$ crystal ($\delta \neq 0$) and the same ratio in the stoichiometric crystal A_nB_m [10]:

$$\Delta = \frac{m+\delta}{n} - \frac{m}{n} = \frac{\delta}{n}. \tag{3}$$

For the three-component system A–B–C, the composition of the solid phase $(A_{1-x}B_x)_{1-y}C_y$ is conveniently expressed in terms of the mole fraction of the binary compound (x) and nonstoichiometry (Δ).

In this case, the nonstoichiometry Δ can be viewed as the difference between the ratio of the equivalent numbers of nonmetal and metal atoms in the real crystal and the same ratio in the stoichiometric crystal. For example, for $(Pb_{1-x}Ge_x)_{1-y}Te_y$ crystals with a NaCl structure,

$$\Delta = y/(1-y) - 1/1 = (2y-1)/(1-y). \tag{4}$$

The mole fraction (molarity) of the binary compound determines the fundamental properties of nonstoichiometric crystals, including the band gap and heat capacity. The concentration of carriers—electrons and holes—and, accordingly, the galvanomagnetic and optical properties of nonstoichiometric crystals are also associated with nonstoichiometry.

3. Directed synthesis of nonstoichiometric solids

The strategy of directed synthesis of substances with the preset nonstoichiometry, structure, and properties is based on physicochemical analysis and includes the steps presented in Figure 2.

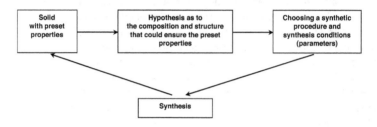

Figure 2. Steps of the directed synthesis of nonstoichiometric solids.

Synthesis is the consequence of processes involved in the conversion of the starting compounds into products. It includes selection and preparation of the starting chemicals (precursors), homogenization of the growth medium (melt, vapor, etc,), the nucleation of the desired phase, nucleus development (growth), and cooling (heating) of the product from the synthesis temperature to room temperature.

Each synthesis step depends on certain conditions or operating parameters. These include the chemical nature and composition of the growth medium, temperature, pressure, diffusion coefficients, heat and mass transfer, and the driving forces of chemical reactions (such as concentration, temperature, pressure, and chemical potential gradients). Figure 3 presents the most important operating parameters of the synthesis.

Provided that the analytical relationship between the rate of solid synthesis (growth) and the operating parameters is known, this process can be carried out under computer control. This problem has been solved for the Czochralski growth of silicon and germanium crystals and for the Bridgman growth of transition metal oxide crystals [10-14].

Solid synthesis involves phase transitions. For this reason, synthesis conditions are selected using phase diagrams (see section 4). The phase diagram of a system indicates the number of compounds forming in this system and the regions of their stability, specifically, the temperature, pressure, and composition intervals. Thus, using the phase diagram, it is possible to choose the medium (melt, vapor, etc.) for the synthesis of the desired solid, the synthesis conditions (temperature, pressure, and growth medium composition), and the way of carrying out the necessary conversions.

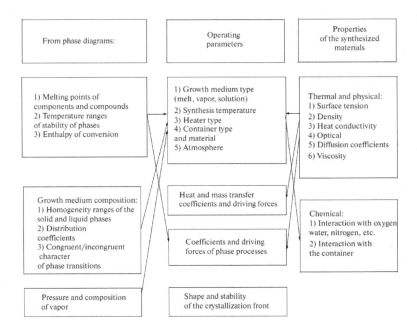

Figure 3. Operating parameters of solid synthesis.

The thermophysical and chemical properties of the starting substances, final synthesis products, and the growth medium to be used (melt, vapor, other solid phases) should be known along with thermodynamic data.

The conversion of a synthesis medium into a solid phase (crystallization) includes the nucleation and development of crystallization centers. Accordingly, it is related with heat and mass transfer and interfacial reaction kinetics. The general problem of analytically describing the crystallization process has not been solved. Furthermore, it is not always possible to evaluate the skill and equipment factors or to establish an unambiguous correlation between the properties of a nonstoichiometric phase and its molecular composition. In this sense, directed synthesis is regarded as being an art [13, 14].

4. Phase diagrams as a key to selecting conditions for synthesis of solid with well-defined nonstoichiometry

The synthetic conditions for nonstoichiometric solids (S_{AB}) coexisting with a vapor (V) and a melt (L) can be estimated using P–T–x diagrams or their T–x projections [15-17]. Let us consider the following features of the T–x diagrams: the highest (maximal) melting point $T_{m,AB}^{max}$ of the solid S_{AB}; the compositions of the phases at this point; some features of the solidus, liquidus, and vapor lines; the nonvariant points of congruent melting (T_m^c), sublimation (T_s^c), and evaporation (T_e^c); and, finally, congruent and incongruent phases and phase processes.

4.1. Maximal melting point $T_{m,AB}^{max}$ of a nonstoichiomertic solid S_{AB}

The Gibbs energy of a phase in a two-component system A–B is given by Eq. (2), so the phase equilibria in this case are represented graphically in a four-dimensional space. This space is explored as four three-dimensional projections: G–P–T, G–P–x, P–T–x, and G–T–x.

To clarify the features of the T–x diagram let us consider the derivation of part of the T–x projection from the G–T–x diagram. For this purpose, it should be considered the relative positions of the G-surfaces of the solid phase S, liquid phase L, and vapor V. In order to determine the arrangement of these three surfaces, we will traverse them with isothermal planes. The projections of the sections of the G-surface on the G–x plane will appear as G^S, G^L, and G^V curves representing the dependence (5) of the Gibbs energy (per gram-atom) on composition at a constant temperature (Figure 4):

$$
\left.
\begin{aligned}
G^S_{A_{1-x}B_x} &= (1-x_B^S)\mu_A^S + x_B^S\mu_B^S \\
G^L_{A_{1-x}B_x} &= (1-x_B^L)\mu_A^L + x_B^L\mu_B^L \\
G^V_{A_{1-x}B_x} &= (1-x_B^V)\mu_A^V + x_B^V\mu_B^V
\end{aligned}
\right\}
\tag{5}
$$

where μ_A^S and μ_B^S are the chemical potentials of the components A and B in the solid phase, μ_A^L and μ_B^L are those of A and B in the liquid phase, μ_A^V and μ_B^V are those of A and B in the vapor phase, and x_B is the mole fraction of the component B.

Let us consider the effect of temperature variation on the relative positions of the G–x curves. At $T_1 > T_{m,AB}^{max}$ (Figure 4a), the solid phase S_{AB} is metastable and the two-phase system L + V is stable because a common tangent line can be drawn for the corresponding G curves, indicating the equality of the chemical potentials of the components in the equilibrium phases.

As the temperature is decreased, the arrangement of the G curves changes. Since the temperature dependence of the Gibbs energy is determined by the entropy $\left(\frac{\partial G}{\partial T}\right)_{P,x} = -S$ and the entropies of the vapor and melt are higher than the entropy of the solid phase, $S^V > S^L > S^S$, the G curves shift upwards upon cooling and the G^V and G^L curves do so more rapidly than the G^S curve. As a consequence, a common tangent for the G^S, G^L, and G^V curves can appear at some temperature $T_2 = T_{m,AB}^{max}$ (Figure 4b). This temperature $T_{m,AB}^{max}$ is referred as the highest maximal melting point of the solid S_{AB}.

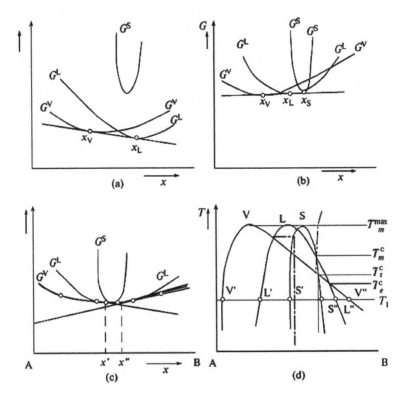

Figure 4. (G–x) sections of the G–T–x diagram: (a) $T_1 > T_{m,AB}^{max}$; (b) $T_2 = T_{m,AB}^{max}$; (c) $T_3 < T_{m,AB}^{max}$; (d) part of the T–x projection near $T_{m,AB}^{max}$.

From the definition of a phase as the totality of the parts of the system whose properties are described by the same equation of state, it follows that the properties of the system are homogeneous functions of composition, pressure, and temperature. Therefore, upon further cooling, for example, to $T_3 < T_{m,AB}^{max}$, it will be possible to draw two tangent lines for each of the G_{AB}^S, G^L, and G^V curves (Figure 4c). The coordinates of the equilibrium phases are designated as V, L, and S in Figure 4d. The subscripts (') and (") are given to the phase compositions to the left and right, respectively, of the composition corresponding to $T < T_{m,AB}^{max}$.

As the temperature changes, the G^V-, G^L- and G^S-curves, shifting upwards and to the sides, describe three surfaces: $G^V = G^V(T, x)$, $G^L = G^L(T, x)$ and $G^S = G^S(T, x)$ – and the projections of the tangency points on the T–x plane draw the solidus line S'–S", the liquidus line L'–L", and the vapor line V'–V" (Figure 4d). Thus, Figure 4d shows part of the T–x projection of the P–T–x diagram near the highest melting point of the S_{AB} compound. The points V', L', and S' and the points V'', L'', and S'', as examples, represent the compositions of the vapor V, solid phase S, and liquid phase L that are in equilibrium at the temperature $T = T_1$ and are called conjugate points. The lines VV'', LL'', and SS'' and the lines VV', LL', SS', as examples, formed by series of these points, are called conjugate lines. The lines connecting conjugate points, such as $V'L'S'$ and $V''L''S''$, are called tie lines. Note that, among the various intersection points between the vapor, liquidus, and solidus lines (Figure 4d), only the intersections points of the conjugate lines (VV'', LL'', SS'', VV', LL', SS') have a physical meaning, as distinct from those of the lines LL' и VV'', etc.

4.2. Specific features of the solids, liquidus and vapor lines

Here, let us consider the continuity of the solidus, liquidus, and vapor lines; the factors determining the homogeneity range; the issue of whether this range must include the stoichiometric composition; the causes of the retrograde character of the solidus line; and the concepts of a pseudocomponent and psevdobinary section in multicomponent systems.

The $T–x$ and $P–T$ projections of the $P–T–x$ diagram of a two-component system having a compound AB [10, 15] are shown in Figure 5. The Gibbs energy of any phase is a homogeneous function of composition, so the solidus line $S'S''$, the liquidus line $L'L''$, and the vapor line $V'V''$, which represent the temperature dependence of the compositions of the equilibrium phases, are continuous lines having no inflection points.

The homogeneity range of a solid compound is bounded by the solidus line. It is determined by the coordinates of the tangency points of the common tangent line for the G curves of the equilibrium phases (Figure 4), i.e., by the equality of the chemical potentials of the components. Therefore, in the general case the homogeneity range (Δ in Eq. (3)) is determined by the relative positions of the G^S, G^L and G^V curves (Figure 4) (i.e., the properties of all coexisting phases) and by the shape of the $(G–x)_{P,T}$ curves. Thus the nonstoichiometry value depends not only the specific features of the nonstoichiometric solid, namely the radius value, electronic configuration and electronegativity of the constituents species, but on the properties of all coexisting phases.

In the case of the compound AB formed from solid components A and B,

$$A_S + B_S = AB_S; \; \Delta_f G^0 \tag{6}$$

the homogeneity range can be estimated using the following relationship [17]:

$$f(\delta' \, / \, k) - f(\delta''/ k) = -\Delta_f G^i, \tag{7}$$

where $\Delta_f G^0$ is the standard Gibbs energy of reaction (6), f is the monotonic function of composition δ, the subscripts ' and " refer to the left and right sides of the homogeneity range, and k is a temperature dependent parameter. It follows from Eq. (7) that, the more negative the Gibbs energy $\Delta_f G^0$, the larger the difference $f(\delta'/k) - f(\delta''/k)$ and, accordingly, the broader the homogeneity range $\delta'' - \delta'$. For crystals dominated by covalent bonding, such as those of the $A^{III} B^{V}$ compounds $GaAs$, InP, etc., $\delta'' - \delta'$ is on the order of thousandths of an atomic percent. For crystal with polar bonds ($CdTe$, $PbSe$, $SnTe$), the homogeneity range is between a few tenths of an atomic percent and several atomic percent.

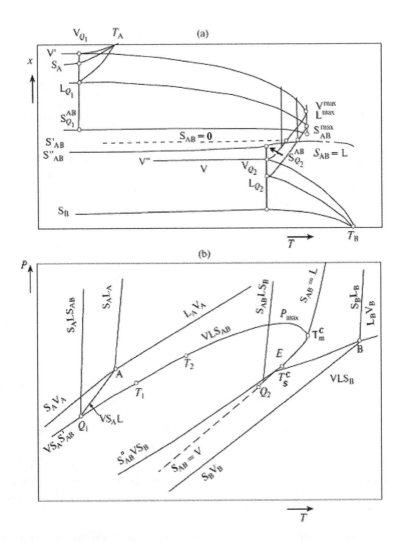

Figure 5. (a) T–x and (b) P–T- projections of the P–T–x diagram of a two-component system with a congruently melting compound (S_{AB}).

The homogeneity range may include the stoichiometric composition or not. Since the $\delta' + \delta''$ value defines the position of the center of the homogeneity range, the expression

$$f(\delta'/k) + f(\delta''/k) = \left[\mu_B(B) - \mu_A(A)\right] / RT - \left[\mu_B(\delta = 0) - \mu_A(\delta = 0)\right] / RT \tag{8}$$

describes the dependence of the position of the homogeneity range center on the difference between the Gibbs energies of the pure components (first term in (8)) relative to the isolated atoms in their ground states and on the difference between the chemical potentials of the components in the stoichiometric solid (second difference in (8). If the latter quantity is neglected in a series of crystallochemically similar compounds, the center of the homogeneity range, $(\delta' + \delta'')/2$ will shift in the direction in which the difference $\mu_B(B) - \mu_A(A)$ increases. This can easily be illustrated by examples of nontransition metal chalcogenides.

In the general case, the stoichiometric composition does not correspond to the minimum of the free energy of the solid phase and can fall outside the homogeneity range. When this is the case, the stoichiometric compound does not exist. For example, strictly stoichiometric ferrous oxide can be obtained only at high pressure.

The temperature dependence of the solidus $S'_{AB}S_{AB}^{max}S''_{AB}$ (Figure 5), which defines the boundaries of the homogeneity range of the nonstoichiometric compound S_{AB}, is described by the following equation [16, 18]:

$$\left(\frac{dx^S}{dT}\right)_P = \frac{(\partial S / \partial x)^S_{P,T} - (S^L - S^S)/(x^L - x^S)}{\left(\dfrac{\partial^2 G}{\partial x^2}\right)_{P,T}} \tag{9}$$

where S^L and S^S are the molar entropies and x^L and x^S are the compositions of the solid phase (S) and melt (L). The solubility of the components in the nonstoichiometric solid $A_{1-x}B_x = =A_{1/2-\delta}B_{1/2+\delta}$ can be retrograde: it can initially increase with increasing temperature $((dx^S/dT)>0)$ and, on passing though a maximum, decrease $((dx^S/dT)_P < 0)$. The maximum solubility $(x^S)_{max}$ is found from the extremum condition for function (9) $((dx^S/dT)_P = 0)$ since $\left(\dfrac{\partial^2 G}{\partial x^2}\right)_P > 0$, according to the phase stability criterion. The retrograde solubility is due to the difference between the rates at which the entropies S^L and S^S, including their configurational components, grow with increasing temperature.

The liquidus line ($L_{Q_1}L^{max}L_{Q_2}$) and vapor line ($V_{Q_1}V^{max}V_{Q_2}$) in Figs. 4 and 5 represent the temperature dependence of the compositions of the melt (L) and vapor (V) in equilibrium with the nonstoichiometric solid S_{AB}.

4.3. Noncoincidence of the phase compositions at the maximum melting point of a nonstoichiometric solid S_{AB}

Both pressure and temperature extrema at T = const and P = const, respectively, exists in a three-phase two-component system provided that the following condition is satisfied [16]:

$$(x^L - x^S)/\left(x^V - x^S\right) = \left(S^L - S^S\right)/\left(S^V - S^S\right) = \left(V^L - V^S\right)/\left(V^V - V^S\right) \tag{10}$$

Thus, if the coexisting phases have different molar volumes V and entropies S, the composition of one of them will be a linear function of the compositions of the two others: $x^L = \beta x^S + (1 - \beta)x^V$ where β is a coefficient depending on the pressure and temperature and independent of the composition. Since $V^L \neq V^S \neq V^V$ and $S^S \neq S^L \neq S^V$ in the general case, at $T = T_{m,AB}^{max}$ the composition of the solid phase S_{AB} does not coincide with the composition of the melt L and vapor V: $x^S \neq x^L \neq x^V$. This is in agreement with the phase rule:

$$c = k + 2 - r - \alpha, \tag{11}$$

where k is the number of components, 2 is the number of external fields (baric and thermal), r is the number of phases, and α is the number of independent constraints imposed on the intensive variables (c). Indeed, if it is assumed that, at $T = T_{m,AB}^{max}$ $x^S = x^L = x^V$, then $\alpha = 2$ in Eq. (11) and $c = 2 + 2 - 3 - 2 = -1 < 0$, which is impossible. Unfortunately, this mistake is frequently encountered in the literature [18-21].

In the case of lead, germanium, tin, and cadmium chalcogenides, the vapor phase consists mainly of a chalcogen, so it can be accepted that $x^S - x^V = -0.5$. At $P \cong 1$ atm, $V^S - V^V \cong 10^{-3} V^S$ and $V^S - V^L \cong 5 \bullet 10^{-2} V^S$. Therefore, $x^S - x^L = 2.5 \times 10^{-5}$; that is, the compositions of the phases at the highest melting point do not coincide, even though they are very similar [14, 22, 23].

4.4. Nonvariant congruent melting, sublimation, and evaporation points of the three-phase equilibrium $S_{AB} + L + V$

Since the solid, liquid, and vapor phases are characterized by different temperature and concentration dependences of the Gibbs energy, the following intersection points of the conjugate liquidus, solidus, and vapor lines can appear on the $T-x$ projection. The temperature of the intersection point of the conjugate liquidus and solidus lines refers to solid and liquid phases of equal compositions ($x^S = x^L$) and is called the congruent melting point T_m^c of the S_{AB} phase. The temperature of the intersection point of the solidus and vapor lines refers to a solid phase and vapor of equal compositions ($x^S = x^V$) and is called the congruent sublimation point T_S^c. The temperature of the intersection point of the conjugate liquidus and vapor lines refers to a liquid phase and vapor of equal compositions ($x^S = x^V$) and is called the congruent evaporation point T_e^c. The temperatures T_m^c, T_S^c and T_e^c do not correspond to the stoichiometric composition ($\delta = 0$) of the solid phase $A_{1/2-\delta} B_{1/2+\delta}$, where $x = 1/2 + \delta$ and δ is the deviation from its stoichiometric composition AB. For example, congruently melting lead telluride contains $(2.8 \pm 0.3)\ 10^{-4}$ mol excess Te per mole of $PbTe$ [22, 23]. Because the formation energies of the defects responsible for the nonstoichiometries $\delta > 0$ and $\delta < 0$ are different, the G^S curve (Figure 4a, 4b, 4c) is asymmetric relative to the $\delta = 0$ composition. Accordingly, the coordinates of the common tangency points of the G curves in the general case do not coincide with the $\delta = 0$ composition.

The equality of the compositions of two phases of the three ones involved in the equilibrium means the appearance of one more relationship $\alpha = 1$ between independent variables (degrees of freedom) in the phase rule expression (11). Therefore, the equilibrium at the points T_m^c, T_S^c and T_e^c is nonvariant: $c = 2+2-3-1 = 0$. The solid S_{AB}, which has a homogeneity range at these points, can be considered a pseudocomponent whose properties are composition-independent, and the sections through ternary, quaternary, and other multicomponent systems involving the S_{AB} solid can be considered as pseudobinary.

4.5. Congruent and inconguent phases and phase processes

The concept of congruence is of great significance in the synthesis of nonstoichiometric solids because, in the case of noncoincidence between the synthetics medium (vapor, melt) and solids compositions, there are fluxes of rejected material and the corresponding kinetic instability of the crystallization front.

A phase can be congruent and incongruent in different temperature intervals. A phase obtainable from the phases that are in equilibrium with it by mixing them in appropriate proportions is called a congruent phase. A phase that cannot be obtained from the coexisting phases is called incongruent [24, 25]. As an example, let us consider the phase S_{AB} (Figure 5a) in the three-phase equilibrium $S_{AB} + L + V$ at different temperatures. Above the congruent melting point ($T_{m,\bullet AB}^c < T < T_{m,AB}^{max}$) the solid phase S_{AB} is incongruent. In the temperature range $T_S^c < T < T_{m,\bullet AB}^c$, it is congruent with respect to the vapor and melt. In designations of three phase equilibria, a congruent phase is written in the middle. Thus, in the former case, the three-phase equilibrium is designated VLS_{AB}; in the latter case, $VS_{AB}L$.

The concepts of congruent and incongruent phases should not be confused with the concepts of congruent and incongruent phase processes.

Phase processes are changes in the state of the system such that the masses of some phases increase owing to the decrease in the masses of others without changes in the intensive parameters (temperature, pressure, phase compositions) [26]. A phase process in which one phase forms or disappears is called congruent. A phase transition in which more than one phase forms or disappears is called incongruent.

The same phase, for example, S_{AB} (Figure 5a), can be involved both in congruent and in incongruent phase processes. At a temperature or pressure corresponding to the three-phase line $Q_1 Q_2$ in Figure 5b, the solid phase S_{AB} melts congruently, yielding a melt and a vapor: $S_{AB} = L + V$. The compositions of the phases involved in this congruent phase process are different and are represented by the solidus, liquidus, and vapor lines in the T–x projection (Figure 5a). The melt and solid phase compositions coincide only at the congruent melting point T_m^c. At the temperatures corresponding to the eutectic non-variant points T_{Q_1} and T_{Q_2} (Figure 5a), the S_{AB} phase is involved in the incongruent phase processes

$$V + S_A = L + S_{AB} \quad \text{and} \quad S_{AB} + L = V + S_B.$$

Thus, S_{AB} phase can be said to melt congruently only in a certain temperature range.

Now let us consider the usage of the terms *congruently melting compound* and *incongruently melting compound* in the literature. Firstly, this differentiation is not strict, because the same compound (e.g., S_{AB}) can be involved in congruent and incongruent phase processes, depending on the temperature. Sometimes, a congruently melting compound is understood as a compound melting without dissociation or decomposition. However, a solid–melt phase transition is accompanied by the breaking and relaxation of chemical bonds in the crystal and by long-range disordering. Therefore, the term *melting without decomposition* is not quite correct, and it should be understood as the identity of the overall compositions of the coexisting phases. In the strict sense, the overall compositions of the phases coincide only at the congruent melting point $T_{m,\bullet AB}^c$, which is below the highest melting point $T_{m,AB}^{max}$: $T_{m,\bullet AB}^c << T_{m,AB}^{max}$.

An incongruently melting compound is sometimes understood as a solid compound that decomposes into a solid phase S_B and a liquid L upon melting (Figure 6). The compositions of the resulting phases differ from the composition of the parent phase. However, in some temperature range, such as $T_{Q_1} < T < T_p$, the "incongruently melting compound" is involved in the congruent melting process $S_{AB} = L + V$.

An essential feature differentiating congruently and incongruently melting compounds is that the highest melting point of a congruently melting compound is higher than the temperatures of the nearest nonvariant points: $T_{m,AB}^{max} > T_{Q_1}$ and T_{Q_2} (Figure 5a).

The highest melting point of an incongruently melting compound is intermediate between the temperatures of the nearest nonvariant points of the system.

Supersaturation and synthesis of a nonstoichiometric solid at a fixed vapor pressure of the volatile component can be produced by cooling or, conversely, heating the three-phase system. The latter case corresponds to the temperature range $T_1 < T < T_2$ in Figure 5b. For example, cadmium telluride crystals were obtained by heating cadmium enriched melts [10]. Note that the composition of the crystals that were grown using the vapor–liquid–crystal technique always lies in the solidus line, i.e., at the boundary of the homogeneity range [15].

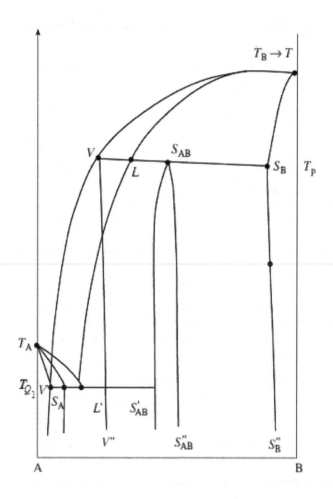

Figure 6. *T–x* projection of the *P–T–x* diagram of a two-component system with an incongruently melting compound (*S*$_{AB}$).

5. Nonstoichiometry and defects in solids

Solids like S_{AB} are grown from the vapor, melt (solution), or solid phases, which are called growth media. Crystallization can be viewed as the transfer of atoms of the components A and B from the growth medium (g.m.) to their regular sites A_A^x and B_B^x in the crystal lattice of the S_{AB} compound:

$$A^{g.m.} = A_A^x + V_B^x; \; \Delta G_1; \quad K_1 = \frac{[A_A^x][V_B^x]}{a_A} = exp(-\Delta G_1 / kT), \tag{12}$$

$$\hat{A}^{g.m.} = B_B^x + V_A^x \; ; \; \Delta G_2; \quad K_2 = \frac{[B_B^x][V_A^x]}{a_B} = exp(-\Delta G_2 / kT) \tag{13}$$

where [ ] designates concentrations and a_A and a_B are the activities of the components in a growth media. The generation of V_B^x and V_A^x vacancies via reactions (12) and (13) is explained by the principle of conservation of the ratio of the numbers of sites characteristic of a given crystal lattice.

Because of the size and energy differences, the Gibbs energies ΔG_1 and ΔG_2, the equilibrium constants of reactions (12) and (13), and, accordingly, the numbers of atoms A and B in the crystals turn out to be different: $K_1 \neq K_2$ and $[A_A^x] \neq [B_B^x]$. Thus, nonstoichiometry appears; that is, the difference between the B-to-A ratios in the real and stoichiometric crystals becomes nonzero.

Note that the properties of crystals are affected not by the nonstoichiometric atoms A_A^x and B_B^x that occupy their regular sites, but by the defects (resulting from a disordering of the ideal structure).

These defects may be the vacancies V_A^x and V_B^x or the interstitial atoms A_i^x and B_i^x

$$A^{g.m.} = A_i^x + \Delta G_3, \tag{14}$$

$$B^{g.m.} = B_i^x + \Delta G_4. \tag{15}$$

This circumstance is due to the fact that the A_A^x and B_B^x species do not change the energy structure of the crystal, but complete it in a way. Near the defects (V_A^x, V_B^x, A_i^x, B_i^x), the energy field and, accordingly, the electrical, mechanical and other properties of the crystal are altered (Figure 7). Thus, defects play an important role in the description of the real structure and properties of nonstoichiometric solids.

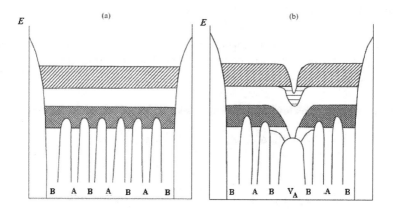

Figure 7. Model of the energy spectrum of (a) an ideal and (b) a nonstoichiometric crystal containing V_A vacancies.

6. Classification and formation of defects

Defect formation processes and defect classification are significant points in control of the defect composition of solids in their directed synthesis.

An ideal, or perfect, solid is one in which all particles making up the substance or structure elements (atoms, ions, molecules, etc.) occupy their regular sites in the lattice. Under heating, irradiation with a beam of high-energy particles, or mechanical treatment, the regular arrangement of particles over their sites undergoes disordering: some particles can leave their sites. The resulting disorder in the arrangement of particles over their normal sites is called defects [27].

In terms of geometry and size, all defects are divided into point and extended defects [27, 28] (Figure 8).

The size of point, or zero-dimensional, defects is comparable with the interatomic parameter. The zero-dimensional (0D) defects include electronic defects (holes, electrons, exitons), energy defects (phonons, polarons), and atomic point defects (APDs). The APDs in nonstoichiometric AB crystals include V_A^x and V_B^x vacancies (absence of atoms or ions in lattice sites); interstitial atoms A_i^x, B_i^x and F_i^x, where F - foreign atoms (the upper symbol x means the neutrality of the defect with respect to its environment), the lower symbol means the defect location.

The formation of an APD is an endothermic process requiring a small amount of energy: $0.5 \leq E_f \leq 3$ eV. Therefore, APDs are equilibrium defects and their concentration depends on synthesis conditions, namely, the temperature and the partial pressures of the components. The size of an APD is 0.1–0.5 nm; however, APDs polarize their environment in the crystal structure, causing slight displacements of neighborions, and largely determine the physical and chemical properties (diffusion, electric conductivity, etc.) of the nonstoichiometric crystal.

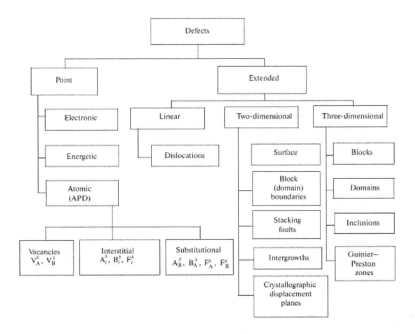

Figure 8. Classification of defects.

Vacancies, interstitial and antistructure (A_B^x , B_A^x) defects are classified as intrinsic defects of a crystal. The concentration (c_j) of these defects takes on a thermodynamically equilibrium value at any $T > 0$ K: $c_j \approx \exp(-W_j/kT)$, where Wj is the defect formation energy, $0 < W_j < 3$ eV. The higher the temperature, the higher the APD concentration c_j. As the crystal is cooled to room temperature, part of the APDs can annihilate via various mechanisms. However, many APDs persist even at exceptionally low reactions (12)–(15), but also from the irradiation of the crystal with high-energy (>1 MeV) electrons at very low temperatures.

Oppositely charged APDs can be attracted to one another to yield new APDs as electroneutral association species, such as ($V_A^- V_B^+$),($V_A^- F_A^+$),(V_A^x)n. Dipole–dipole interaction leads to the formation of APD clusters, which can serve as nuclei of other phase in the nonstoichiometric crystal.

Extended defects include linear (1D), or one-dimensional, and two (2D)- and three-dimensional (3D) defects [29, 30] (Figure 8). Let us consider some specific features of these defects.

Linear defects, or dislocations, are similar to point and two-dimensional defects in the sense that their size is comparable with the unit cell parameter. In the third dimension, the dislocations are fairly long or even infinite. The simplest kind of dislocation is the edge dislocation.

The most important two-dimensional (2D) defects include solid surface, block (domain) boundaries, stadling faults and crystallographic displacement planes, which are surfaces in which coordination polyhedra of two contacting ideal are rearranged.

An example of periodic intergrowths is provided by the family of tungsten bronzes, A_xWO_3 (A = alkali metal). In these compounds, WO_3 layers are intergrown with hexagonal bronze bands. The formation of intergrown bronzes is likely favored by the synthesis conditions. The boundaries of intergrowths can be periodic (ordered) or nonperiodic (disordered).

Note the following specific features of two-dimensional defects. Firstly, their formation energy is fairly high (>3 eV) and they are kinetically stabilized nonequilibrium defects. The "frozen" state of these defects is responsible for their "memory" for preparation history. Secondly, these defects do not change the stoichiometry of the substance. Thirdly, planar defects result from APD interaction and exert a significant effect on on the reactivity and physical properties of nonstoichiometric solids.

The size of three-dimentional defects (Figure 8) exceeds the lattice constant in all three directions. These defects are, in essence, macroscopic imperfections of the crystal structure and are formed during crystal growth and subsequent processing. Three-dimensional defects include separate blocks; mosaics (totality of a large number of small- and large-angle boundaries); inclusions (microdeposits) resulting from phase transitions, such as the decomposition of a solid solution; magnetic domains (crystal zones with the same orientation of spins or electric dipoles); Guinier–Preston zones (parallel platelike formations as thick as a few unit cells, separated by different distances and having the same composition as the crystal); cavities; and cracks. Three-dimentional defects can be viewed as resulting from defect association and ordering. For example, pores can be considered to result from the association of a large number of vacancies. Bulk defects also include elastic tensile and compressive stresses.

Deviations from stoichiomrtry may be so large that defect interactions become significant leading to direct ordering, clustering, superstructure formation, long-range ordering, and the formation of new nonstoichiometric phases differing in symmetry, energy and other aspects from the parent phase. In such systems defects are intrinsic components of the crystal structure rather than being statistically distributed imperfections. The crystal-chemical and thermodynamic aspects of nonstoichiometric compounds with narrow and broad homogeneity range as well the approaches for controlling the nonstoichiometry and considered in [31].

7. Substance homogeneity criterion in physicochemical analysis

The Gibbs energy (G) of a crystal is a statistical quantity related to the distribution function. The mean G value determines the most likely distribution of zero-, one-, two-, and three-dimensional defects. Fluctuations around the mean value are possible, and, in a closed system (solid substance with a constant net composition at a fixed temperature), the configurational fluctuation can manifest itself as a change in composition within a small region as a consequence of a random motion of particles inside some volume element or to its surface. This brings up the question of what substance can be called homogeneous. The degree of heterogeneity of a solid phase is characterized by the statistical distribution of structure elements from which the crystal is built. These structure elements are the atoms of the components of the system in their regular sites, as well as various zero-, one-, two-, and three-dimensional defects in their regular positions (Figure 8).

A quantitative estimate of the degree of heterogeneity can be based on the following three types of distributions:

1. distribution of structure elements in some measurable volume,

2. distribution of these volumes in the crystal, and

3. distribution of measurement data and properties of the solid phase.

Let σ be the confidence interval and c_i be the concentration of structure elements in the ith microvolume. If

$$\left| c_i - 1/N \sum_{i=1}^{N} c_i \right| \leq \sigma \tag{16}$$

the solid substance can be called homogeneous. If there is an i value at which this inequality is invalid, the solid phase should be considered to be heterogeneous. For practical use of a material, of significance are such deviations of the property in a given object from the weighted average value for the entire system that go beyond the confidence interval. In this sense, heterogeneity can be understood as the totality of property values, measured in all microvolumes, that fall outside the confidence interval. An analysis of generalized heterogeneity criteria in terms of an autocorrelation function was carried out by Nikitina et al. [32]. A thermodynamic analysis of defect ordering and an interpretation of the concentration dependences of physical properties taking into account short- and long- range order parameters were made using a model based on the above stricture elements or cluster components [33].

Some features of homogeneity criteria, physicochemical analysis, thermodynamics in materials science and inorganic crystal engineering are discussed in [34-37].

8. Conclusion

Technological progress is intimately associated with creation of new materials, such as composites, piezoelectrics, ferroelectrics, semiconductors, superconductors and nanomaterials with preset functional properties. For the synthesis of these materials, it is necessary to study the chemical processes that lead to the desired properties

Physicochemical analysis is the field of chemistry dealing with these processes, the interrelation between the composition, structure and properties of matter and determination of synthetic conditions for obtaining such substances. Several challenges and new approaches have been discussed. They include the concepts of a substance, phase, component, directed synthesis strategy as well as some definitions. Attention was paid to nonstoichiometry, classification and formation of defects.

Synthesis of nonstoichiometric solid involves control over phase transformations. The P–T–x phase diagram is a key to selecting conditions for synthesis of solid with well-defined nonstoichiometry. For this reason the following features of P–T–x phase diagrams were considered: highest melting point of a nonstoichiometric compound ($T^{max}_{m,AB}$), noncoincidence between the solid-, liquid-, and vapor-phase compositions ($x^L \neq x^S \neq x^V$) at this temperature, factors determining the nonstoichiometry range, the terms congruent and incongruent phases and phase processes. Criteria for evaluating the degree of heterogeneity of nonstoichiometric solids were also considered.

Glossary of Abbreviations

$T^{max}_{m,AB}$ maximal melting point of a nonstoichiometric compound AB

(Na^+_{Na}) cationic sites

(Cl^-_{Cl}) anionic sites

U internal energy

Q heat

A work execution

δ nonstoichiometry

T^c_m nonvariant point of congruent melting

T^c_S nonvariant point of congruent sublimation

T^c_e nonvariant point of congruent evaporation

S solid phase

L liquid phase

V vapor

μ^S_A chemical potential of the component A in solid phase

μ^S_B chemical potential of the component B in solid phase

μ^L_A chemical potential of the component A in liquid phase

μ^L_B chemical potential of the component B in liquid phase

μ_A^V chemical potential of the component A in vapor phase

μ_B^V chemical potential of the component B in vapor phase

x_B mole fraction of the component B

S^V entropy of vapor

S^L entropy of melt

S^S entropy of solid

G^V Gibbs energy of vapor

G^L Gibbs energy of liquid

G^S Gibbs energy of solid

$\Delta_f G^\circ$ standard formation Gibbs energy

k number of components

r number of phases

α number of independent constraints imposed on the intensive variables (c)

A_A^x regular sites of atom A in crystal lattice of the S_{AB}

B_B^x regular sites of atom B in crystal lattice of the S_{AB}

K_1 equilibrium constant

K_2 equilibrium constant

V_B^x and V_A^x vacancies

A_i^x and B_i^x interstitial atoms

c_j concentration

W_j defect formation energy

σ confidence interval

$(V_A^- V_B^+)$ electroneutral association species

$(V_A^- F_A^+)$ electroneutral association species

Author details

V. P. Zlomanov[1], A.M. Khoviv[2] and A.Ju. Zavrazhnov[2]

*Address all correspondence to: zlomanov@inorg.chem.msu.ru

2 Department of Chemistry, Voronezh State University, Voronezh, Russia
1 Department of Chemistry, Moscow State University, Moscow, Russia

References

[1] Kurnakov N.S. Zadachi Inst. Physikochemical Analysis. Izv. Inst. Phys.Chem. Analysis. 1919; 1(1), 1-7.

[2] Kurnakov N.S. The Introduction to Physicochemical Analysis. 4th ed. Moscow: Akad. Nauk SSSR; 1940 [in Russian].

[3] Kurnakov N.S. Collection of Selected Works. Leningrad/Moscow: ORTI, Khimteoret; 1938; vol. II, p. 1-202 [in Russian].

[4] Kurnakov N.S. Selected Works. Moscow:Akad. Nauk SSSR; 1963; vol. II, 1-361 [in Russian].

[5] Anosov V.Ya., Pogodin S.A. The Principles of Physicochemical Analysis. Moscow: Akad.Nauk SSSR; 1947 [in Russian].

[6] Tamman G. The Manual on Heterogeneous Equilibria. Leningrad:ONTI, Khimteoret; 1935 [inRussian].

[7] Gibbs J.W. Thermodynamics: Statistical Mechanics. Moscow: Nauka; 1982.

[8] Voronin G.F. The Fundamentals of Thermodynamics. Moscow:Moskow State University; 1987 [in Russian].

[9] Fuks G.I. The Smallest Piece, the Smallest Drop. Khimya and Zhizn'. 1984; 2, 74-76 [in Russian].

[10] Kröger F,A. The Chemistry of Imperfect Crystals. Amsterdam: North-Holland Publ. Co.; 1973.

[11] Vasiliev Ya., Akhmetshin R., Borodlev Yu. BGO Crystal Growth by a Low Thermal Gradient Czochralski Technique. Nucl. Instr. Meth. Phys. Res. 1996; A379, 533.

[12] Golyshev V., Gonic M. et al. Heat Transfer in Growing $Be_4Ge_3O_{12}$ Crystals under Weak Convection. J. Cryst. Growth. 2004; 262, 202-214.

[13] Rosenberger R.F. Fundam. Cryst. Growth. New-York: Springer-Verlag; 1979.

[14] Gottshtain G. The Physical Foundation of Materials Science. Berlin-Heidelberg: Springer-Verlag; 2004.

[15] Zlomanov V.P., Novoselova A.V. P–T–x Diagrams of Metal–Chalcogen Systems. Moscow: Nauka; 1987 [in Russian].

[16] Storonkin A.V. The Thermodynamcis of Heterogeneous Systems. Leningrad: Leningrad State University; 1967; part 1 [in Russian].

[17] Brebrick R.F. Nonstoichiometry in Binary Semiconductor Compounds $M_{0.5-\delta}N_{0.5+\delta}$. Prog. Solid State Chem. 1966; 3, 213-263.

[18] Roozeboom W.B. Die heterogenen Gleichgewichte vom Standpunkte der Phasenlehre. Braunschweig: Vieweg; 1904.

[19] Ricci J.E. The Phase Rule and Heterogeneous Equilibrium. New York: Dover; 1966.

[20] Khaldoyanidi K.A. Phase Diagrams of Heterogeneous Systems. Novosibirsk: INKh RAN; 1991; part 1 [in Russian].

[21] Levinskii Yu.V. p–T–x Diagrams of Binary Metal Systems. Moscow: Metallurgiya; 1990; book 1 [in Russian].

[22] Peter K., Wenzel A., Rudolph P. The $p-T-x$ Projection of the System $Cd-Te$. Cryst. Res. Technol. 1990; 25(10), 1107-1114.

[23] Avetisov I. Kh., Mel'kov A.Yu., Zinov'ev A.Yu., Zharikov E.V. Growth of Nonstoichiometric PbTe Crystals by the Vertical Bridgman Method Using the Axial-Vibration Control Technique. Crystallography Reports. 2005; 50, Suppl. 1, 124-129.

[24] Meuerhaffer W., Sanders A.Z. Z. für Phys.Chem. IV. Phys. Chem. 1899; 28, 453.

[25] W. Meuerhaffer W., Sanders A.Z. Z. für Phys.Chem. II. Phys. Chem. 1905; 53, 513-518.

[26] Münster A. Chemische Thermodynamik. Berlin: Academie – Verlag; 1969.

[27] Van Bueren H.G. Imperfections in Crystals. Amsterdam: North-Holland; 1960.

[28] Fistul' V.I. The Physics and Chemistry of the Solid State. Moscow: Metallurgiya; 1995 [in Russian].

[29] West A. Solid State Chemistry and Its Applications. Chichester: Wiley; 1984.

[30] Rabenau A., editor. Problems of Nonstoichiometry. Amsterdam: North Holland; 1970.

[31] Zlomanov V.P. Crystal Growth of Nonstoichiometric Compounds. Inorganic Materials. 2006, 42, Suppl. 1, 19-48.

[32] Nikitina V.G., Orlov A.G., Romanenko V.N. Problem of Nonhomogeneity Distribution of Atoms and Defects in Semiconductor Crystals. In: Growth of Semiconductor Crystals and Films. Novosibirsk: Nauka; 1981; 204 [in Russian].

[33] Men' A.N., Bogdanovich M.P., Vorob'ev Yu.P. et al. Composition–Imperfection–Properties of Solid Phases. Cluster Components Method. Moscow: Nauka; 1977 [in Russian].

[34] De Hoff Robert T. Thermodynamics in Materials Science. Taylor & Francis. Boca Raton. London. New-York. 2011. P. 531.

[35] Brammer L. Developments in Inorganic Crystal Engineering. Chem. Soc. Rev. 2004. V.33. P. 476-489.

[36] Shriver & Atkins. Inorganic Chemistry. Oxford University Press. Fourth Ed. 2006. 822 P.

[37] Ashby M., Sherclift H., Ceban D. Materials, Engineering, Science, Processing and Design. Elsevier. Amsterdam. Boston. 2011.

Chapter 6

Nanocrystalline Mn and Fe Doped ZnO Thin Films Prepared Using SILAR Method for Dilute Magnetic Semiconductor Application

Rathinam Chandramohan,
Jagannathan Thirumalai and
Thirukonda Anandhamoorthy Vijayan

Additional information is available at the end of the chapter

1. Introduction

Zinc oxide (ZnO) is a versatile material of compound semiconductors with excellent properties and extensive applications in electronics, optoelectronics, sensors, and catalyses (Das et
al, 2007). ZnO thin films have attracted considerable attention because they can be tailored
to possess high electrical conductivity, high infrared reflectance and high visible transmittance by different coating technique (Ryu et al, 2000). Some of the remarkable properties of
ZnO are its wide direct-band gap of 3.37 eV, the binding energy of the exciton of ZnO (60
meV) which makes it an excellent material for excitonic devices (Wang et al, 2003). Other
favourable aspects of ZnO include its broad chemistry leading to many opportunities for
wet chemical etching, low power threshold for optical pumping, radiation hardness and bio-
compatibility. Together, these properties of ZnO make it an ideal candidate for a variety of
devices ranging from sensors through to ultra-violet laser devices and nanotechnology
based devices such as displays. As fervent research into ZnO continues, difficulties such as
the fabrication of p-type ZnO that have so far stated that the development of devices had
overcome (Yang etal, 2008). Mitra et al (1998) has prepared Zinc Oxide thin films using
chemical deposition technique. The structural, morphological properties of the prepared
films are characterized using X-ray diffraction and scanning electron microscope. They have
used Zn salts as precursor and successfully synthesized ZnO films. The growth of highly
textured Zinc oxide (ZnO) thin films with a preferred (101) orientation has been prepared by

employing chemical bath deposition using a sodium zincate bath on glass substrates has been reported by (Ramamoorthy et al, 2004).

(Natsume et al, 2000) have studied the d.c electrical conductivity and optical properties of zinc oxide film prepared by a sol-gel spin coating technique. The temperature dependence of the conductivity indicated that electron transport in the conduction band was due to thermal execution of donor electrons for temperatures from 250 to 300 K. (Chapparro et al, 2003) have proposed the spontaneous growth of ZnO thin films from aqueous solutions. An electroless – chemical process is proposed, consisting in the formation of the super oxide radical (O_2^-) followed by chemical reaction of two O_2^- with Zn $(NH_3)_4^{2+}$ cations. (Wellings et al, 2008) have deposited ZnO thin films from aqueous zinc nitrate solution at 80ºC onto fluorine doped tin oxide (FTO) coated glass substrates. Structural analysis, surface morphology, optical studies and electrical conductivity were studied and thickness of the ZnO films was found to be 0.40 μm. (Walter et al, 2007) have studied the characterization of strontium doped ZnO thin films on love wave filter applications. X-ray diffraction, scanning electron microscopy and atomic force microscopy studied the crystalline structure and surface morphology of films. The electrochemical coupling coefficient, dielectric constant, and temperature coefficient of frequency of filters were then determined using a network analyzer. (Vijayan et al, 2008, a, b; Chandramohan et al, 2010) have reported the preparation conditions for undoped ZnO using double dip technique and used them for gas sensor applications. They have also reported the synthesis of Sr doped ZnO using double dip technique and used them for gas sensor applications. Recently (Chandramohan et al, 2010) have synthesized Mg doped ZnO thin films using double dip chemical growth and reported the ferromagnetic properties of the films. Saeed et al. (1995) have deposited thin films of mono phase crystalline hexagonal ZnO from solutions of zinc acetate in the presence of ethylenediamine and sodium hydroxide on to glass microscope studies. Two distinct morphologies of ZnO were observed by scanning electron microscopy. The deposited films were specular and adherent. (Cheng et al 2006) have fabricated thin films transistors (TFTs) with active channel layers of zinc oxide using a low – temperature chemical bath deposition. Current voltage (I-V) properties measured through the gate reveal that the ZnO channel is n-type. (Sadrnezhaad et al 2006) have studied the effect of addition of Tiron as a surfactant on the microstructure of chemically deposited zinc oxide. Addition of tiron charges the surface morphology and causes to form the fine – grained structure. The obtained results indicate that increasing the number of dipping carves to progress the deposition process. (Piticescu, et al 2007) have studied the influence of the synthesis parameters on the chemical and microstructural characteristics of nanophases synthesized in the two methods. 'Al' doping tends to a lower material density and to a smaller gown size. Zhou et al (2007) have studied microstructure electrical and optical properties of aluminium doped zinc oxide films. The ZnO:Al thin films are transparent (~ 90%) in near ultraviolet and visible region A. with the annealing temperature increasing from 300ºC to 500ºC. The film was oriented more preferentially along the (002) direction, the grain size of the film increased, the transmittance also became higher and the electrical resistivity decreased. Bulk ZnO is quite expensive and unavailable in large wafers. So, for the time being, thin films of ZnO are relatively a good choice. Usually, the doped ZnO films with optimum properties (perfect crystalline structure, good conducting properties, high transparency, high intensity of luminescence) are obtained when they are grown on heated substrates and annealed after deposition at high temperature in oxygen atmosphere (Peiro et al, 2005 Lo-khande et al, 2000; Srinivasan et al, 2006; Chou et al, 2005). However, for an extensive use in the commercial applications pure and doped ZnO films must be prepared at much lower substrate temperatures. Therefore, it is necessary to develop a low-temperature deposition technology for the growth of ZnO films. Many works are seen in the low temperature growth of this interesting ZnO system both undoped and metal doped (Tang et al, 1998, Cracium et al, 1994; Gorla et al, 1999; Kotlyarchuk et al, 2005) thin films and nano thin films.

Fe addition is expected to enhance the sensing ability of this system. When sensing of 300 ppm ethanol, the in situ extended X-ray absorption fine structure (EXAFS) spectra indicate that the bond distances of Zn-O and Fe-O are 1.90 and 1.98 Å, respectively and restored to 1.91 and 1.97 Å in the absence of ethanol (Hsu et al., 2007). The single-phase Fe-doped ZnO films ($x \leq 0.04$) exhibit ferromagnetism above room temperature (Wei et al., 2006). The Fe/ZnO composites display photoluminescence properties different from those of the ZnO entities generated similarly but in the absence of Fe powder: (i) a UV emission band is ob-served over the latter but not over the former, and (ii) the former shows an emission peak around 583 nm (wavelength) with intensity significantly stronger than that of the latter. Be-cause of the encapsulation of the Fe nanoparticles, the Fe/ZnO composites are highly stable in air and high magnetization (Yang et al., 2009). Room temperature ferromagnetic Fe:ZnO film was prepared by chemical deposition of ZnO film and chemical introduction of Fe im-purity in aqueous solutions. (Nielsen, et al., 2008) have reported on Magnetization meas-urements show clear ferromagnetic behavior of the magnetite layers with a saturation magnetization of $3.2 \mu_B$/f.u. at 300 K. Their results demonstrate that the Fe3O4/ZnO system is an intriguing and promising candidate for the realization of multifunctional heterostruc-tures. During recent years different techniques such as chemical vapor deposition (Hu, et al., 1991), magnetron sputtering (Jiang, et al., 2003), spray pyrolysis (Pawar, etal., 2005), sol-gel (Cheong et al., 2002), and molecular beam epitaxy (Kato et. al., 2002), have been used for deposition of zinc oxide films on various substrates. The ease with which the deposi-tion can be made and the reproducibility of the method to produce doped oxide films in any laboratory make this method quite attractive (Kato et. al., 2002) (Vijayan et al, 2008, a, b) (Chandramohan et al, 2010). While the sol-gel method with double dipping has pro-duced highly oriented thin films of Li, Mg doped ZnO thin films by (Liu et al., 2005), they have introduced a spin in the dipped films using spin coating set up to spread the films over substrate. In the method used in this work no spinning is required as the dehydrogen-ation of the films is done using a dip in hot water. The magnetic properties of the metal doped ZnO thin films grown by such a method has not been presented in detail to our knowledge. Nanostructured ZnO not only possesses high surface area, nontoxicity, good bio-compatibility and chemical stability, but also shows biomimeticand high electron com-munication features. A reagentless uric acid biosensor based on ZnO nanorodes can be syn-thesized from SILAR synthesized ZnO thin films. The enzyme is immobilized on ZnO nanocombs and nanorods to construct an amperometric biosensor for glucose biosensing. Besides such systems have prospective applications in Dye-sensitized solar cells (DSSCs), Sensor, optoelectronic devices, UV detector, Spintronics, Dilute Magnetic Semiconductor devices etc. SILAR is a modification of the chemical bath deposition technique. In CBD the solution is in contact with a substrate for a longer period and the reaction is slow and the formation is good (Hodes et. al., 2002). Accelerated by Bath temperature, agitation or bath conditions. The film formation results when ionic product exceeds solubility product. In SILAR originally called as multiple dip method by (Ristov, et al., 1987) and named SILAR (Nicolau, et al., 1988) the reaction Cationic and anionic precursor solutions separated for reaction at chosen conditions like temperature, pH, dipping rate, waiting time, number of dipping cycles. Time is important in ionic layer formations. Since deposition is in air too many microstructural variations are possible. The aim of this work is to investigate the influence of the preparation conditions on structural, optical and electrical properties of ZnO and incorporate magnetic properties by doping Mn/Fe leading to dilute magnetic semi conducting system using CBD method. Advantages are effectiveness and simplicity of the deposition equipment, high deposition rates, wide spectrum of deposition parameters for the control and the optimization of film properties, and film thickness. The sum of all these special features enables the growth of oxide thin films at low temperature substrates with perfect crystallinity. The present work is a preparation and characterization of undoped ZnO, Mn- doped ZnO (MZO) and Fe- doped ZnO (FZO) thin films by chemical deposition technique. In which the influence of solution concentration, solution pH value, film thickenss, annealing temperature and concentration of strontium and aluminium atoms of the grown films are investigated. In addition it demonstrates that any dopant can be used in principle along with the precursor to enable them to be included in the system. The technique can be tuned to get the desired morphology and nanocrystallites of desired sizes distributed over any type of substrate for various applications.

2. Synthesis, structure and morphology of the ZnO thin films

2.1. Synthesis

ZnO thin films were grown using a two-step chemical bath deposition technique using a solution comprising 0.1 M Zinc Sulphate (99% emerk), 0.2 M sodium hydroxide with a pH value of 9±0.2 deposited at bath temperature of 90 °C under optimized condition. For Mg doped ZnO (MZO) thin films $Fe(SO4)_3$ was used at a concentration of 0.1mM. Before deposition, the glass substrates were cleaned by chromic acid followed by cleaning with acetone. The well-cleaned substrates were immersed in the chemical bath for a known standardized time followed by immersion in hot water for the same time for hydrogenation. The process of solution dip (step 1) followed by hot water dipping (step 2) is repeated for known number of times. The cleaned substrate was alternatively dipped for a predetermined period in sodium zincate bath and water bath kept at room temperature and near boiling point, respectively. According to the following equation, the complex layer deposited on the substrate during the dipping in sodium zincate bath will be decomposed to ZnO due to dipping in hot water. The proposed reaction mechanism is according to the following equations (Vijayan et al, 2008, a, b) (Chandramohan et al, 2010).

$$ZnSO_4 + 2\,NaOH \rightarrow Na_2ZnO_2 + H_2SO_4 \uparrow$$

$$Na_2ZnO_2 + H_2O \rightarrow ZnO + 2\,NaOH$$

Part of the ZnO so formed was deposited onto the substrate as a strongly adherent film and the remainder formed as a precipitate. The addition of MSO_4 in the ratio of Zn:M as 100:1 in the first dip solution leads to the formation of Mg/Fe doped zinc oxide nano thin films where M stands for Mg/Fe.

2.2. Structural analysis

The crystallographic structure of the films has been studied by X-ray diffraction (XRD). Fig. 1 shows the XRD spectrum of ZnO, MZO and FZO films deposited on the glass substrate under optimized condition. It can be seen from the XRD data, that all samples are polycrystalline and exhibit the single-phase ZnO hexagonal wurtzite structure [JCPDS (36–1451)]. All peaks in recorded range were identified.

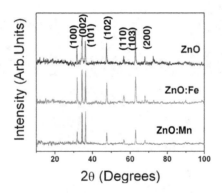

Figure 1. Typical XRD patterns of doped and undoped ZnO thin films

The XRD pattern clearly showed the polycrystalline nature of the ZnO, MZO and FZO films, whose c-axis was preferentially oriented normal to the glass substrate. In other words, those grains of undoped and doped films are mainly grown with c-axis vertical to the glass substrate. Hence, the multiple-coating or the piling up of each film was considered not to disturb the overall growth of the films with c-axis orientation. Therefore, the c-axis orientation may be a common phenomenon in the ZnO film deposition by the chemical process using organo-zinc compounds. Such preferred basal orientation is typically observed also in metal doped ZnO films (Vijayan et al, 2008, a, b) (Chandramohan et al, 2010). Moreover, from the recorded spectrums the minor diffraction peaks of (102) and (103) are approved of randomly oriented of the ZnO film (Roy, at al., 2004). The crystallite size was estimated to be 80 nm for undoped film, 26 nm for MZO and 20 nm for FZO films from the Debye Scherrer formula.

2.3. Morphological studies

Figures 2(a, b, c) shows the scanning electron micrograph (SEM) of ZnO, MZO and FZO films deposited at room temperature. The SEM micrograph of MZO thin film show the uniform polycrystalline surface of the film with a hexagonal morphology consistent with XRD result of P63mc crystal-structure with an average grain size of 300 nm. They are found to be single-crystalline in nature. It can be seen that, films grown at room temperature by varying Mn and Fe concentration consist of slightly agglomerated particles with less voids in the surface of the film with average grains 200 nm for undoped and 300 nm for doped ZnO films (Fig. 2(b)). This result confirmed that Mn and Fe doped into ZnO lattice and in good agreement line with XRD result.

Figure 2. (a,b,c,) Typical SEM micrographs obtained for (a) undoped ZnO thin films and (b) Mn (2%), (c) Fe (2%) doped thin films prepared by SILAR

3. Optical properties of undoped and Fe doped ZnO thin films

Figure 3 shows the transmittance spectrum of the MZO, FZO thin films grown on glass substrate. ZnO is a non-stoichiometric oxide and is known to contain zinc-ion excess defects based on the presence of either zinc interstitial or oxygen vacancies. The films have excellent transmittance and very low absorption and reflectance. The optical band gap of the FZO and MZO thin films has decreased on Fe and Mn doping, respectively. Assuming doping levels are well below Mott's critical density, the change in optical band gap can be explained in terms of Burstein–Moss band gap widening and band gap narrowing due to the electron–electron and electron–impurity scattering. At high doping concentrations, fermi level lifts into the conduction band. Due to filling of the conduction band, absorption transitions occurs between valance band and fermi level in the conduction band instead of valance band and bottom of the conduction band. This change in the absorption energy levels shifts the absorption edge to higher energies (blue shift) and leads to the energy band broadening. While on Mn and Fe doping into the ZnO matrix can explain the increase in shift in the band gap value indicating that either it may due to any charged defects or the charged defects formed had been neutralized by other defects. Hence, the blue shift in the band gap value by Mn and Fe doping suggest an increase in the n-type carrier concentration, most of the Fe ions must be incorporated as interstitial donors into the structure rather than substitution of acceptors.

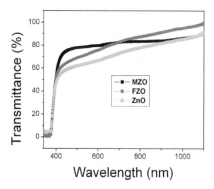

Figure 3. Optical transmittance of typical doped and undoped ZnO thin films

4. Magnetic properties of undoped and Mn, doped ZnO thin films

The ZnO thin films when studied using VSM show a Coercivity (Hci) 238.70G Under Field at Ms/20 G, its Hci, Negative -230.81 G and Hci, Positive 246.60 G, respectively. The undoped sample had Magnetization (Ms) 3.0830E-3 mu/cm^2 and Retentivity (Mr) 368.68E-6 emu/cm^2 respectively. The field at ms/2 0G specifies the absence of any soft or hard magnetic property. The magnetization M-H curve (Figure 4) also shows a behaviour anaalogus to a non magnetic material for undoped ZnO thin films. However the FZO and MZO sample showed the characteristics of a ferromagnetic behaviour magnetic field, and the saturated magnetization (Ms) is 697.32E-6 emu/cm^2. It's a well known fact that the magnetic properties of dilute magnetic systems can be explained by bound magnetic polaron (BMP) model (Hu, et al., 1992; Jiang, et al., 2003). According to this model an impurity site (donor or acceptor) plays the role of a trap and captures the carrier (electrons or holes) to form a bound polaron. These polarons are usually surrounded by the magnetic Mg ions. The polaron interaction with magnetic Mg ions causes the alignment fully or partially to generate the magnetic property of the system. Considering the morphology of the Metal doped ZnO films which had Zn defects and Fe$^+$/Mn$^+$ in the ZnO:Mn film. The ferromagnetism observed in the film can be explained by using the BMPs model. The magnetic exchange interaction between ZnO and Fe$^+$/Mn$^+$ occupying the same space is aligned with Fe^{1+}/Mn^{1+} spins, forming BMPs. Thus, the sample can exhibit ferromagnetism. Similar behavior of room temperature ferromagnetism is exhibited by Mn doped ZnO thin films grown by SILAR.

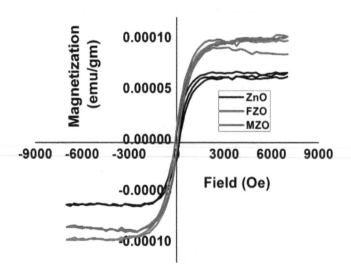

Figure 4. (a) The VSM spectra of undoped and Fe and Mn doped ZnO thin films grown by SILAR.

In summary, Polycrystalline, hexagonal ZnO, MZO and FZO thin films with (002) preferential orientation have been deposited from aqueous solutions using a modified two-step chemical bath deposition technique onto a glass substrate. The microstructure of the films are studied and reported. The studies revealed the potential of this SILAR method in creating and designing varieties of morphologies suitable for various applications. Optical absorption indicated the shift in band gap to 3.21 and 3.22 eV respectively for FZO and MZO films with respect to band gap of ZnO matrix and refractive index to be around 2.34 for FZO films and the bandgap decreased to 3.22 eV for MZO films with a refractive index around 2.3. The transmittance became higher for both MZO and FZO films with increase in doping concentration. In the doped ZnO films, the films were oriented more preferentially along the

(002) direction, the grain size of the films decreased, the transmittance also became higher and the electrical resistivities decreased. It is shown that doped ZnO thin films deposited with a CBD technique can have high temperature ferromagnetic property and this transition may be explained by the BMP model. Extensive characterizations on the structure, microstructure optical and electrical properties have been made and the exotic choice available in this simple method has paved way for the synthesis of many similar systems by our group like Fe, Mg and Mn doped ZnO thin films and other TCO systems like CdO, etc. Also the properties of these thin film nanocrystallites can be tailored to suit variety of applications like, phosphors, display panels, thermal conduction and opto electronic devices. The technique is easy for automation and anticorrosive coatings can be coated employing doped ZnO systems on to various mechanical spares. The potential of this technique is yet to be exploited in full by the industrial community. The crystallite shape and size control is also feasible in this excellent method.

Acknowledgements

Dr.R.C. Thanks UGC, New Delhi for supporting part of this work through a major research project. Also Dr. R.C thanks Dr. P. Parameshwaran PMD, MSG, IGCAR, Kalpakkam India and C. Gopalakrishnan, Department of Nanotechnolgy, SRM University, Chennai extending SEM facilities.

Author details

Rathinam Chandramohan[1], Jagannathan Thirumalai[2] and
Thirukonda Anandhamoorthy Vijayan[2]

1 Department of Physics,Sree Sevugan Annamalai College,Devakottai, Tamil Nadu, India

2 Department of Physics, B.S. Abdur Rahman University, Vandalur, Chennai, Tamil Nadu, India

References

[1] Chandramohan, R.; Thirumalai, J.; Vijayan, T. A.; ElhilVizhian, S.; Srikanth, S.; Valanarasu, S. & Swaminathan, V. (2010). Nanocrystalline Mg Doped ZnO Dilute Magnetic Semiconductor Prepared by Chemical Route. Adv. Sci. Lett. 3., 3., (September & 2010) 319-322, ISSN: 1936-6612.

[2] Chapparro, M.; Maffiotte, C.; Gutierrez, M.T. & Herrero, J. (2003). Study of the spontaneous growth of ZnO thin films from aqueous solutions.Thin solid films., 431., 1., (May & 2003) 373-377, ISSN: 0040-6090.

[3] Cheong, K.Y.; Muti, N.; Ramanan, S.R. (2002). Electrical and optical studies of ZnO:Ga thin films fabricated via the sol–gel technique. Thin Solid Films. 410., 1-2., (May 2002), 142-146., ISSN: 0040-6090.

[4] Chou, T– L.; Ting, & J– M. (2005). Deposition and characterization of a novel integrated ZnO nanorods/thin film structure. Thin solid films. 494., 1-2., (January & 2006) 291-295, ISSN: 0040-6090.

[5] Cracium, V.; Elders, J.; Gardeniers, J.G.E. & Boyd, L. W. (1994). Characteristics of high quality ZnO thin films deposited by pulsed laser deposition. Appl. Phys. Lett., 65., 23., (October & 1994) 2963-2965., ISSN: 0003-6951.

[6] Das, S.; Chaudhuri, S. (2007). Mg^{2+} substitutions in $ZnO–Al_2O_3$ thin films and its effect on the optical absorption spectra of the nanocomposite. Appl. Surf. Sci., 253., 21., 8661-8668., ISSN: 0169-4332.

[7] Feng, Z–C.; Chen, C–F.; Kuo, C-T; Williams, K.; Shan, W. The 3rd Asian Conference on Chemical Vapor Deposition (3rd Asian-CVD), Taipei, Taiwan, November 12–14, 2004. Thin solid films., 498., 1-2., (March & 2006) 1, ISSN: 0040-6090.

[8] Gorla, C.R.; Emanetoglu, N.W.; Liang, S.; Mayo, W.E.; Cu, Y.; Wraback, M. & Shen, H. (1999). Structural, optical, and surface acoustic wave properties of epitaxial ZnO films grown on (011-2) sapphire by metalorganic chemical vapor deposition. J. Appl. Phys., 85., 5., 2595-2602., ISSN 0021-8979.

[9] Hodes, G.; Chemical Solution Deposition of Semiconductor Films, Marcel Dekker Inc., Oct. 2002.

[10] Hsu, H.H.; Paul Wang, H.; Chen, C.Y.; Jou C.J.G.; Wei, Y-L. (2007). Chemical structure of zinc in the Fe/ZnO thin films during sensing of ethanol, J. Elect. Spec. Rel Phen., 156-158., (December & 2007), 344-346., ISSN: 0368-2048.

[11] Hu, J.; Gordon, R. G. (1992). Textured aluminum-doped zinc oxide thin films from atmospheric pressure chemical-vapor deposition. J. Appl. Phys. 71., 2., (October & 1991), 880-891, ISSN 0021-8979.

[12] Hu, J.; Gordon, R.G.; (1992). Textured aluminum-doped zinc oxide thin films from atmospheric pressure chemical-vapor deposition. J. Appl. Phys. 71., 2., (October 1991), 880-890, ISSN 0021-8979.

[13] J. Cryst. Growth 237–239., Part 1., (December & 2001), 538-543., ISSN: 0022-0248.

[14] J. Cryst. Growth. 92., 1-2., (October & 1988) 128-142., ISSN: 0022-0248.

[15] J. S. Wellings, A. P. Samantilleke, P. Warren, S. N. Heavens and I. M. Dharmadasa. (2008). Comparison of electrodeposited and sputtered intrinsic and aluminium-doped zinc oxide thin films. Semicond. Sci. Technol. 23., 12., (August & 2008) 125003-125009, ISSN 0268-1242.

[16] Jiang, X.; Wong, F.L.; Fung, M.K.; Lee, S.T. (2003). Aluminum-doped zinc oxide films as transparent conductive electrode for organic light-emitting devices. Appl. Phys. Lett. 83., (July & 2003), 1875-1877, ISSN: 0003-6951.

[17] Jiang, X.; Wong, F.L.; Fung, M.K.; Lee, S.T.; (2003). Aluminum-doped zinc oxide films as transparent conductive electrode for organic light-emitting devices. Appl. Phys. Lett. 83., 9., (July & 2003), 1875-1877., ISSN 0003-6951.

[18] Kato, H.; Sano, M.; Miyamoto, K.; Yao, T. (2002). Growth and characterization of Ga-doped ZnO layers on a-plane sapphire substrates grown by molecular beam epitaxy.

[19] Kotlyarchuk, B.; Sarchuk, V. & Oszwaldowski, M. (2005). Preparation of undoped and indium doped ZnO thin films by pulsed laser deposition method. Cryst. Res. Technol., 40., 12., (December & 2005) 1118- 1123, ISSN: 0232-1300.

[20] Liu, C.; Yun, F.; & Morkoç, H, (2005). Ferromagnetism of ZnO and GaN: A Review J. Mater. Sci: Mat in Electronics. 16., 9., 555-597, ISSN: 0957-4522.

[21] Lokhande, B.J. & Uplane, M.D. (2000). Structural, optical and electrical studies on spray deposited highly oriented ZnO films. Appl. Surf. Sci., 167., 3-4., (October & 2000) 243-246, ISSN: 0169-4332.

[22] Mitra, P.; Chatterjee, A.P. & Maiti, H.S. (1998). Chemical deposition of ZnO films for gas sensors. J. Mater. Sci: Mat in Electronics., 9., 6., (December & 1998) 441-445, ISSN: 0957-4522.

[23] Natsume, Y. & Sakata, H. (2000). Zinc oxide films prepared by sol-gel spin-coating. Thin solid films., 372., 1-2., (September & 2000) 30-36, ISSN: 0040-6090.

[24] Nicolau, Y.F.; Menard, J.C.,; (1988). Solution growth of ZnS, CdS and $Zn_{1-x}CdxS$ thin films by the successive ionic-layer adsorption and reaction process growth mechanism.

[25] Nielsen, A.; Brandlmaier, A.; Althammer, M.; Kaiser, W.; Opel, M.; Simon, J.; Mader, W.; Goennenwein, S.T.B.; Gross, R. (2008). All Oxide Ferromagnet/Semiconductor Epitaxial Heterostructures. Appl. Phys. Lett. 93., 16., (October & 2008) 162510-3., ISSN: 0003-6951.

[26] Pawar, B.N.; Jadkar, S.R.; Takwale, M.G. (2005). Deposition and characterization of transparent and conductive sprayed ZnO:B thin films. J. Phys. Chem. Solids. 66., 10., (November & 2005) 1779-1782., ISSN: 0022-3697.

[27] Peiro, A. M.; Ayllon, J. A.; Pearl, J.; Domenech, X. & Domingo C. (2005). Microwave activated chemical bath deposition (MW-CBD) of zinc oxide: Influence of bath composition and substrate characteristics. J. Crystal. Growth., 285., 1-2., (November &2005) 6-16, ISSN: 0022-0248.

[28] Piticescu, R. R. Piticescu, R. M. & Monty, C. J. (2006). Synthesis of Al-doped ZnO nanomaterials with controlled luminescence. J. Europ. Cer. Soc. 26., 14., (March & 2006) 2979-2983, ISSN: 0955-2219.

[29] Ramamoorthy, K.; Arivanandhan, M.; Sankaranarayanan, K. & Sanjeeviraja, C. (2004). Mater. Chem. Phys., 85., 2-3., (June & 2004) 257-262, ISSN: 0254-0584.

[30] Ristov, M.; Sinadinovski, G.; Grozdanov, I.; Mitreski, M. (1987).Chemical deposition of ZnO films. Thin Solid Films. 149., 1., (May & 1987), 65-71., ISSN: 0040-6090.

[31] Roy, V. A. L.; Djurisic, A. B.; Liu, H.; Zhang, X. X.; Leung, Y. H.; Xie, M. H.; Gao, J.; Lui, H. F.; Surya, C. (2004). Magnetic properties of Mn doped ZnO tetrapod structures. Appl. Phys. Lett. 84, 5., (2004), 756-759., ISSN: 0003-6951.

[32] Ryu, Y. R.; Zhu, S.; Budai, J. D.; Chandrasekhar, H. R.; Miceli, P. F.; White, H.W. (2000). Optical and structural properties of ZnO films deposited on GaAs by pulsed laser deposition. J. Appl. Phys. 88., 1., 201-204, ISSN: 0021-4922.

[33] Sadrnezhaad, S.K. & Vaezi, M.R. (2006). The effect of addition of Tiron as a surfactant on the microstructure of chemically deposited zinc oxide. Mat. Sci. Engg: B., 128., 1-3., (March & 2006) 53-57, ISSN: 0921-5107.

[34] Saeed, T. & Brien, P. O'. (1995). Deposition and characterisation of ZnO thin films grown by chemical bath deposition. Thin solid films., 271., 1-2., (December & 1995) 35-38, ISSN: 0040-6090.

[35] Srinivasan, G. & Kumar, J. (2006). Optical and structural characterisation of zinc oxide thin films prepared by sol-gel process. Cryst. Res. Technol. 41., 9., (September & 2006) 893- 896, ISSN: 0232-1300.

[36] Tang, Z.K.; Wang, G.K. L.; Yu, P.; Kawaraki, M.; Ohtomo, A.; Koinuma, H. & Segawa, Y. (1998). Appl. Phys. Lett. 72., 3270, ISSN: 0003-6951.

[37] Vijayan, T. A.; Chandramohan, R.; Valanarasu, S.; Thirumalai, J.; Venkateswaran, S.; Mahalingam, T. & Srikumar, S.R. (2008). Optimization of growth conditions of ZnO nano thin films by chemical double dip technique. Sci. Tech. Adv. Mater., 9., (April & 2008) 035007, ISSN: 1468-6996.

[38] Vijayan, T. A.; Chandramohan, R.; Valanarasu, S.; Thirumalai, J.; Subramanian, S. P. (2008). Comparative investigation on nanocrystal structure, optical, and electrical properties of ZnO and Sr-doped ZnO thin films using chemical bath deposition method. J. Mater. Sci., 43., 6., (March & 2008) 1776–1782, ISSN: 0022-2461.

[39] Vijayan, T. A.; Chandramohan, R.; Valanarasu, S.; Thirumalai, J.; Subramanian, S. P. (2008). Comparative investigation on nanocrystal structure, optical, and electrical properties of ZnO and Sr-doped ZnO thin films using chemical bath deposition method. J. Mater. Sci., 43., 6., (March & 2008) 1776–1782, ISSN: 0022-2461.

[40] Wang, L.; Giles, N. C. (2003). Temperature dependence of the free-exciton transition energy in zinc oxide by photoluminescence excitation spectroscopy. J. Appl. Phys. 94., 2., 973-978., ISSN: 0021-4922.

[41] Wei, X. X.; C Song, K W Geng, F Zeng, B He and F Pan. (2006). Local Fe structure and ferromagnetism in Fe-doped ZnO films. J. Phys.: Condens. Matter. 18., 31., (July & 2006) 7471-7479., ISSN 0953-8984.

[42] Yang, Y.; Tay, B.K.; Sun, X.W.; Han, Z.J.; Shen, Z.X.; Lincoln, C.; & Smith, T, (2008). Nanoelectronics Conference, INEC 2008 2 nd IEEE International, Nanyang Technical University, Singapore 24-27 March 2008.

[43] Yang, Z. X.; Zhong, W.; Au, C. T.; Du, X.; Song, H. A.; Qi, X. S.; Ye, X. J.; Xu M. H.; Du, Y. W. (2009). Novel Photoluminescence Properties of Magnetic Fe/ZnO Composites: Self-Assembled ZnO Nanospikes on Fe Nanoparticles Fabricated by Hydrothermal Method. J. Phys. Chem. C., 113., 51., (November & 2009) 21269 – 21273., ISSN: 1932-7447.

[44] Zhou, H–M.; Yi, D–Q.; Yu, Z–M.; Rang, L.; Xiao, Li, J. (2007). Preparation of aluminum doped zinc oxide films and the study of their microstructure, electrical and optical properties. Thin solid films., 515., 17., (June 2007) 6909-6914, ISSN: 0040-6090.

Chapter 7

Mechanochemical Synthesis of Magnetite/Hydroxyapatite Nanocomposites for Hyperthermia

Tomohiro Iwasaki

Additional information is available at the end of the chapter

1. Introduction

Hydroxyapatite ($Ca_{10}(PO_4)_6(OH)_2$, HA), which is a calcium phosphate ceramic, has been widely used as a biomaterial in various applications (e.g., artificial bone and dental root, cosmetic foundation, etc.) because of its high biocompatibility and chemical stability. Moreover, many attempts are being made to give new functions to HA by incorporating effective components into a HA matrix. In particular, magnetite (Fe_3O_4)-incorporated HA (Fe_3O_4/HA) nanocomposites have attracted much attention as a promising material for hyperthermia therapy of malignant bone tumor [1–4]. Recently, Fe_3O_4/HA composites have also been used as adsorbents [5–7] and catalysts [8,9].

Fe_3O_4/HA composites can be synthesized conventionally by mixing HA powder with Fe_3O_4 nanoparticles which are prepared individually [1–3,5–10]. The conventional synthesis methods have disadvantages: reaction time required for completing the formation of HA and Fe_3O_4 is relatively long, subsequent heat treatments for long periods of time are required for aging and crystallization. Thus, the synthesis of Fe_3O_4/HA composites generally consist of multi-step processes. Therefore, a simple method which can provide Fe_3O_4/HA composites rapidly is needed to be developed.

In this chapter, a mechanochemical method for the simple synthesis of Fe_3O_4/HA nanocomposites is presented. In this method, superparamagnetic Fe_3O_4 nanoparticles are first prepared mechanochemically from ferric hydroxide [11], and then the mechanochemical synthesis of HA from dicalcium phosphate dihydrate ($CaHPO_4 \cdot 2H_2O$) and calcium carbonate ($CaCO_3$, calcite) is performed [12–14], followed by the aging for a short period of time. These mechanochemical treatments are sequentially performed in a single horizontal tumbling ball mill at room temperature under wet conditions. The wet mechanochemical process can also contribute to the distribution of Fe_3O_4 nanoparticles in the HA matrix, which can result in a good hyperthermia property. In addition, the use of horizontal tumbling ball mills is reasonable for the synthesis of Fe_3O_4/HA nanocomposites because the device structure is simple, the handling is easy, the energy consumption is relatively low, and the scale-up is easy [15]. The influence of conditions on the formation of Fe_3O_4/HA nanocomposites was investigated and the hyperthermia property was examined. The details are described below.

2. Mechanochemical synthesis of hydroxyapatite nanoparticles

First of all, the synthesis of HA nanoparticles containing no Fe_3O_4 nanoparticles was investigated to optimize the synthesis process of Fe_3O_4/HA nanocomposites. In all the experiments presented in this chapter, the chemicals of analytical grade were used as received without further purification. Typically, 30 mmol of $CaHPO_4$ $2H_2O$ and 20 mmol of $CaCO_3$, corresponding to the stoichiometric molar ratio in the formation reaction of HA expressed by Equation (1) [14], were added to 60 ml of deionized and deoxygenated water.

$$6CaHPO_4 \cdot 2H_2O + 4CaCO_3 \rightarrow Ca_{10}\left(PO_4\right)_6\left(OH\right)_2 + 14H_2O + 4CO_2 \qquad (1)$$

The resulting suspension was subjected to a mechanochemical treatment using a horizontal tumbling ball mill, as illustrated in Figure 1. The suspension was placed in a Teflon-lined milling pot with an inner diameter of 90 mm and a capacity of 500 ml. Zirconia balls with a diameter of 3 mm were used as the milling media; the charged volume of the balls (including voids among the balls) was 40% of the pot capacity. The wet milling was performed at room temperature in air atmosphere under atmospheric pressure for a designated period of time. The rotational speed was 140 rpm, corresponding to the ideal critical rotational speed. After milling, the precipitate was isolated from the suspension by centrifugation, washed with acetone, and dried at room temperature in air. As a control experiment without milling, the starting suspension was vigorously stirred at room temperature for 24 h.

The samples obtained under various conditions were characterized according to standard methods. The powder X-ray diffraction (XRD) pattern of samples was obtained by an X-ray diffractometer (RINT-1500, Rigaku; CuKα radiation, 40 kV, 80 mA, $2\theta=5°–50°$, scanning rate: $1.0°/min$). Figure 2 shows the XRD pattern of samples obtained in different milling times. As the milling time increased, the diffractions indicating the presence of $CaHPO_4$ $2H_2O$ and $CaCO_3$ decreased. Simultaneously, the diffractions indicating HA appeared. In particular, a drastic change was observed between 1 h and 3 h. On the contrary, when stirred for 24 h without milling, the XRD pattern (not shown) hardly changed from the beginning, which was almost the same as that before milling as shown in Figure 2a. These results indicate that the milling promoted the solid phase reaction expressed by Equation (1). However, after milling for 12 h, the XRD pattern was almost the same and the diffraction at 2θ=29.4°, indicating the presence of $CaCO_3$, still remained even in 24 h.

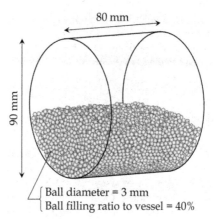

Figure 1. Schematic illustration of horizontal tumbling ball mill used in this work.

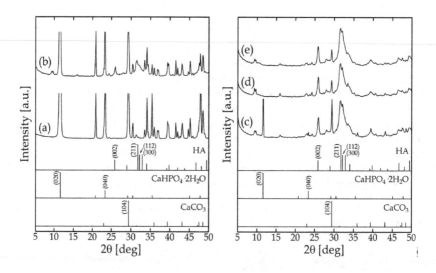

Figure 2. XRD pattern of samples (a) before milling and after milling for (b) 1 h, (c) 3 h, (d) 12 h, and (e) 24 h.

The differential scanning calorimetry (DSC) was performed using a thermal analyzer (SDT2960, TA Instrument) with an argon flow rate of 100 ml/min. The temperature was raised from ambient temperature to 900 °C at a rate of 20°C/min. Figure 3 shows the results of DSC analysis for the raw materials and the samples. In the sample obtained in 1 h (Figure 3d), the endothermic peaks were clearly observed at around 200°C and 750°C, which resulted from the elimination of water of crystallization in CaHPO$_4$ 2H$_2$O and the thermal decomposition of CaHPO$_4$ 2H$_2$O and CaCO$_3$. Although the peaks relating to CaHPO$_4$ 2H$_2$O disappeared as the milling time, the peak resulted from the thermal decomposition of CaCO$_3$ remained even in 12 h. Accordingly, it was found that the milling was not sufficient to complete the formation reaction of HA.

The morphology of samples was observed by field emission scanning electron microscopy (FE-SEM; JSM-6700F, JEOL). Figure 4 shows typical SEM images of samples. In a milling time of 1 h, coarse particles coated with fine particles of about 100 nm were observed. From the particle size analysis of CaHPO$_4$ 2H$_2$O and CaCO$_3$ by the laser diffraction/scattering method (SALD-7100, Shimadzu), the median sizes were determined to be 16.2 μm for CaHPO$_4$ 2H$_2$O and 2.0 μm for CaCO$_3$. In general, horizontal tumbling ball mills are difficult to produce nanoparticles for short milling times. Therefore, coarse and fine particles could be the raw materials and HA, respectively. As the milling time increased, coarse particles disappeared and the number of HA nanoparticles increased. However, even after 12 h, a little number of coarse particles was found.

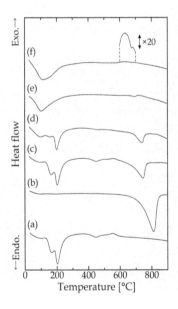

Figure 3. DSC curve of (a) CaHPO$_4$ 2H$_2$O, (b) CaCO$_3$, and samples (c) before milling and after milling for (d) 1 h, (e) 3 h, and (f) 12 h.

In order to complete the formation reaction of HA, the heat treatment (aging) was performed after milling. For investigating the effect of heating on the formation of HA, the un-milled suspension of $CaHPO_4 2H_2O$ and $CaCO_3$ was heated under various conditions of temperature and time. Figures 5, 6, and 7 show the XRD patterns of samples obtained without milling after heating at 40, 60, and 80°C, respectively. When the suspension was heated at 40°C, the formation reaction of HA hardly took place. As increasing in the temperature, the reaction was promoted and could complete at 80°C in 8 h. Thus, when without milling, higher heating temperatures and longer heating times are needed for the formation of HA.

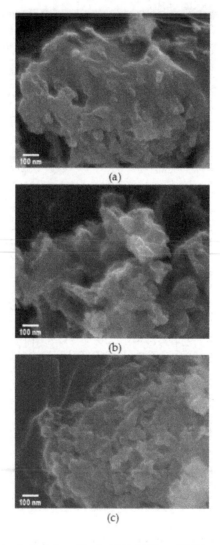

(a)

(b)

(c)

Figure 4. SEM image of samples obtained after milling for (a) 1 h, (b) 3 h, and (c) 12 h.

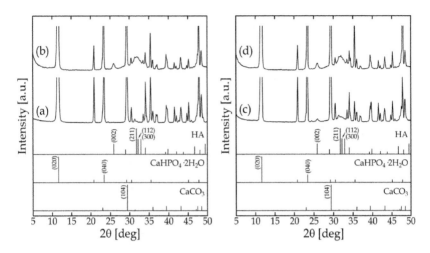

Figure 5. XRD pattern of un-milled samples (a) before heating and after heating at 40°C for (b) 3 h, (c) 5 h, and (d) 8 h.

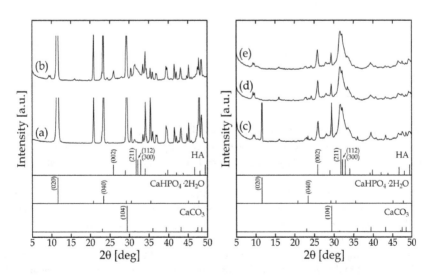

Figure 6. XRD pattern of un-milled samples (a) before heating and after heating at 60°C for (b) 1 h, (c) 3 h, (d) 5 h, and (e) 8 h.

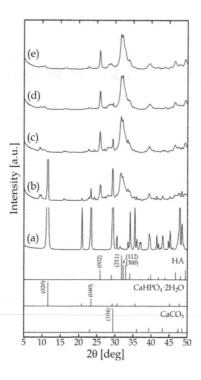

Figure 7. XRD pattern of un-milled samples (a) before heating and after heating at 80°C for (b) 1 h, (c) 3 h, (d) 5 h, and (e) 8 h.

Next, the effect of milling of the suspension before heating on the formation of HA was investigated. Figures 8–14 show the XRD patterns of samples obtained under various conditions of milling time, heating temperature, and heating time. It was found that longer milling times, higher heating temperatures, and longer heating times promoted the formation reaction of HA. In particular, as shown in Figure 10c, when the heating was performed at 80°C, only the milling for 1 h and the following heating for 1 h provided a single phase of HA. The SEM images of samples obtained by milling for different times under constant heating conditions of 80°C and 1 h are shown in Figure 15. When heating at 80°C for 1 h, a typical morphology of HA was observed regardless of the milling time. However, the particle size intended to decrease as the milling time increased. Consequently, the combination of milling and heating of the suspension of $CaHPO_4$ $2H_2O$ and $CaCO_3$ can produce efficiently HA for short periods of time.

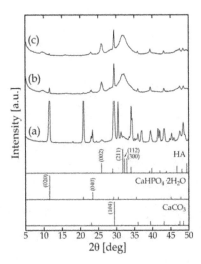

Figure 8. XRD pattern of 1 h-milled samples (a) before heating and after heating at 40°C for (b) 1 h, and (c) 5 h.

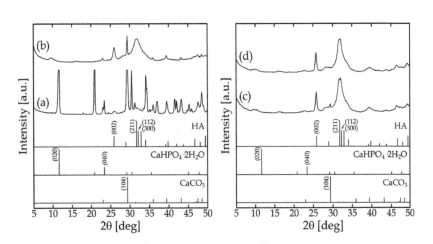

Figure 9. XRD pattern of 1 h-milled samples (a) before heating and after heating at 60°C for (b) 1 h, (c) 3 h, and (d) 5 h.

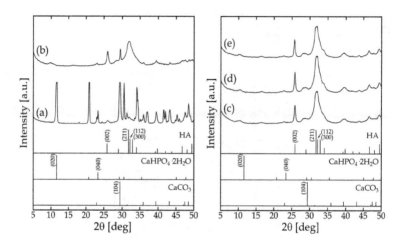

Figure 10. XRD pattern of 1 h-milled samples (a) before heating and after heating at 80°C for (b) 30 min, (c) 1 h, (d) 3 h, and (e) 5 h.

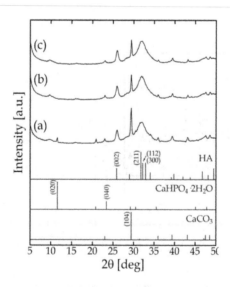

Figure 11. XRD pattern of 3 h-milled samples (a) before heating and after heating at 40°C for (b) 1 h, and (c) 5 h.

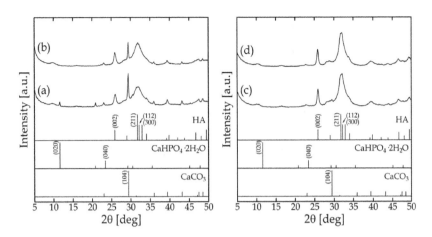

Figure 12. XRD pattern of 3 h-milled samples (a) before heating and after heating at 60°C for (b) 1 h, (c) 3 h, and (d) 5 h.

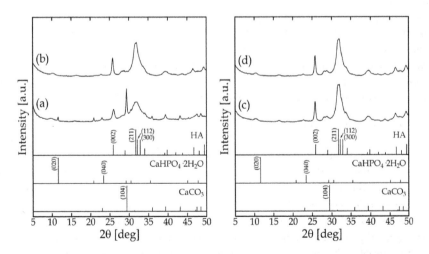

Figure 13. XRD pattern of 3 h-milled samples (a) before heating and after heating at 80°C for (b) 1 h, (c) 3 h, and (d) 5 h.

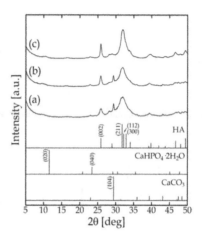

Figure 14. XRD pattern of 12 h-milled samples (a) before heating and after heating at 80°C for (b) 30 min, and (c) 1 h.

3. Synthesis and hyperthermia property of magnetite/hydroxyapatite nanocomposites

In the synthesis of Fe_3O_4/HA nanocomposites, first a suspension of superparamagnetic Fe_3O_4 nanoparticles was prepared according to a mechanochemical method reported in elsewhere [11]. This method provides Fe_3O_4 from ferric hydroxide (goethite) in the absence of a reducing agent; goethite is reduced to ferrous hydroxide by mechanochemical effects and the solid phase reaction between ferrous hydroxide and goethite generates Fe_3O_4 [16]. Subsequently, HA nanoparticles were synthesized in the suspension of Fe_3O_4 nanoparticles in the same container by the mechanochemical method mentioned above.

4.5 mmol of ferric chloride hexahydrate ($FeCl_3$ $6H_2O$) was dissolved in 60 ml of deionized and deoxygenated water. To precipitate amorphous ferric hydroxides (mostly, goethite), a proper amount of 1.0 M sodium hydroxide (NaOH) solution was dropped into the solution which was magnetically stirred under a continuous flow of argon at room temperature. The pH was adjusted to higher than 13. A brown suspension thus prepared was placed in a gas-tight milling pot (inner diameter 90 mm, capacity 500 ml) made of 18%Cr–8%Ni stainless steel. Stainless steel balls (diameter 3.2 mm) were used as the milling media. The charged volume including the voids among the balls was about 40% of the pot capacity. The pot was purged of air, filled with argon, and sealed. The milling was performed at room temperature for 11 h. The rotational speed was 140 rpm, corresponding to the ideal critical rotational speed.

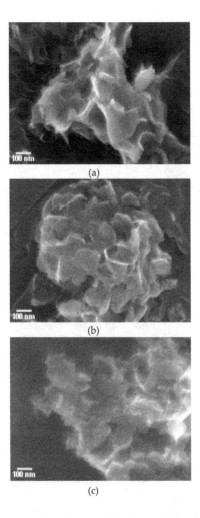

(a)

(b)

(c)

Figure 15. SEM image of samples obtained by milling for (a) 1 h, (b) 3 h, and (c) 12 h, followed by heating at 80°C for 1 h.

The XRD pattern of Fe_3O_4 nanoparticles thus prepared is shown in Figure 16. The Fe_3O_4 nanoparticles had a high crystallinity and an average crystallite size of 11.7 nm which was calculated from the full width at half-maximum (FWHM) of the Fe_3O_4 (311) diffraction peak at $2\theta=35.5°$ using Scherrer's formula. The lattice constant was determined to be 8.387 Å from several diffraction angles showing high intensity peaks, which was close to the standard value of Fe_3O_4 (8.396 Å) as compared to that of maghemite (8.345 Å). Figure 16 also shows that no reflections indicating formation of other compounds were observed. This indicates the Fe_3O_4 nanoparticles were high purity.

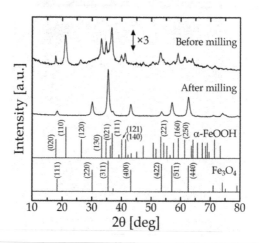

Figure 16. XRD pattern of Fe₃O₄ nanoparticles prepared by mechanochemical method.

As shown in Figure 17, the SEM image indicated that the Fe_3O_4 nanoparticles had a diameter of approximately 10–20 nm, which almost agreed with the average crystallite size (11.7 nm). The hydrodynamic size (number basis) was measured by dynamic light scattering (DLS; Zetasizer Nano ZS, Malvern Instruments) for a dispersion, as shown in Figure 18. The median diameter was determined to be 16.4 nm from the size distribution, which was also near the average crystallite size. These results reveal that the Fe_3O_4 nanoparticles have a single-crystal structure.

The magnetic property (magnetization-magnetic field hysteretic cycle) was analyzed using a superconducting quantum interference device (SQUID) magnetometer (Quantum Design model MPMS) at room temperature in the rage of magnetic field between –10 kOe and 10 kOe. Figure 19 shows the magnetization-magnetic field curve. The Fe_3O_4 nanoparticles had a low coercivity (4 Oe), showing superparamagnetism. The saturation magnetization (78 emu/g) was a little lower than that of the corresponding bulk (=92 emu/g) because of the smaller size [17].

Figure 17. SEM image of Fe₃O₄ nanoparticles prepared by mechanochemical method.

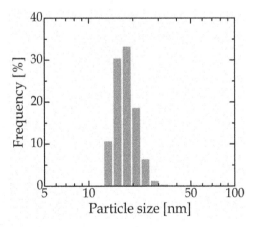

Figure 18. DLS particle size distribution of Fe₃O₄ nanoparticles prepared by mechanochemical method.

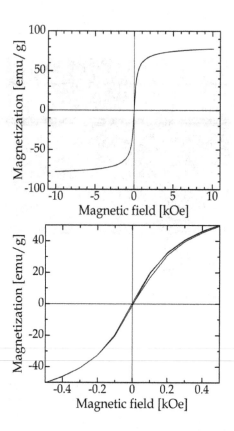

Figure 19. Magnetization-magnetic field curve of Fe_3O_4 nanoparticles prepared by mechanochemical method.

After the suspension of Fe_3O_4 nanoparticles was prepared, the milling pot was opened, and then predetermined amounts of $CaHPO_4$ $2H_2O$ and $CaCO_3$ were added to the suspension. Their amounts were adjusted so that the mass concentration of Fe_3O_4 nanoparticles in the Fe_3O_4/HA nanocomposite was 10, 20, and 30 mass%. In order to prevent the oxidation of Fe_3O_4 during milling, the pot was purged of air, filled with argon, and sealed prior to milling. The suspension was milled at a rotational speed of 140 rpm for 1 h at room temperature, followed by the heating at 80°C for 1 h.

Figure 20 shows the XRD pattern of Fe_3O_4/HA nanocomposites with different Fe_3O_4 concentrations. It was confirmed that the nanocomposites consisted of Fe_3O_4 and HA having no by-products regardless of the Fe_3O_4 concentration. The average crystallite sizes of Fe_3O_4 and HA were calculated from the FWHM of the Fe_3O_4 (311) plane at $2\theta=35.5°$ and the HA (002) plane at $2\theta=25.9°$, respectively, using Scherrer's formula, and listed in Table 1. The average crystallite sizes of Fe_3O_4 and HA were almost constant regardless of the concentration of Fe_3O_4 in the Fe_3O_4/HA nanocomposites.

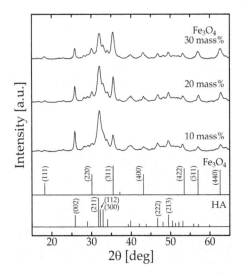

Figure 20. XRD pattern of Fe₃O₄/HA nanocomposites with different Fe₃O₄ concentrations.

Fe$_3$O$_4$ concentration	Crystallite size of Fe$_3$O$_4$	Crystallite size of HA
10 mass%	11.3 nm	20.3 nm
20 mass%	12.8 nm	18.8 nm
30 mass%	9.8 nm	17.8 nm

Table 1. Crystallite sizes of Fe₃O₄ and HA in Fe₃O₄/HA nanocomposites.

Figure 21 shows the SEM image of nanocomposite containing 30 mass% Fe₃O₄ as an example. The Fe₃O₄ nanoparticles with a diameter of about 20 nm were distributed homogeneously in the HA matrix without forming large aggregates. It was confirmed that nanometer-sized Fe₃O₄/HA composite particles were successfully synthesized.

The magnetic hyperthermia property was evaluated using an apparatus reported in elsewhere [18]. A proper amount of Fe₃O₄/HA nanocomposite powder sample was placed in a polystyrene tube with a diameter of 16 mm, and packed by tapping the tube. The packing volume was constant at 0.8 cm³ regardless of the Fe₃O₄ concentration. The temperature increase was measured in an AC-magnetic field using an optical fiber thermometer. The frequency and amplitude of the AC-magnetic field were 600 kHz and 2.9 kA/m, respectively. Figure 22 shows the temperature increase for the nanocomposites in the AC-magnetic field. As the Fe₃O₄ concentration increased, the temperature increased more rapidly. When the Fe₃O₄ concentration was 30 mass%, the temperature increase of 40°C was achieved only after about 20 sec. This result supports that the Fe_3O_4/HA nanocomposites synthesized by this mechanochemical process exhibit a good hyperthermia property [1–4].

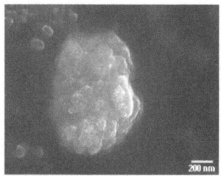

Low-magnification observation

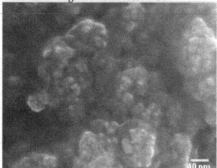

High-magnification observation

Figure 21. SEM images of 30 mass% Fe_3O_4/HA nanocomposites.

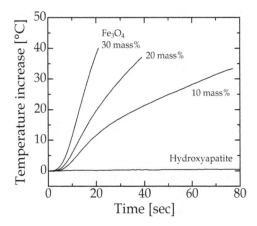

Figure 22. Temperature profiles of HA nanoparticles and Fe_3O_4/HA nanocomposites with different Fe_3O_4 concentrations in AC-magnetic field.

4. Conclusion

A mechanochemical method for the simple synthesis of Fe_3O_4/HA nanocomposites has been developed, in which superparamagnetic Fe_3O_4 nanoparticles and HA nanoparticles are sequentially prepared in a single horizontal tumbling ball mill at room temperature under wet conditions. First, the synthesis process of HA (containing no Fe_3O_4) was optimized. The obtained HA samples were characterized by XRD, DSC, and SEM. The influence of conditions on the formation of HA nanoparticles was investigated. Mechanochemical effects induced during wet milling promoted the reactions between $CaHPO_4$ $2H_2O$ and $CaCO_3$ forming HA even at room temperature. The combination of milling and heating (aging) of the suspension of $CaHPO_4$ $2H_2O$ and $CaCO_3$ can produce efficiently HA for short periods of time. The optimum operating conditions in the synthesis of HA were determined as follows: a rotational speed of 140 rpm, a milling time of 1 h, an aging temperature of 80°C, and an aging time of 1 h. Next, the synthesis of Fe_3O_4/HA nanocomposites was investigated. The mechanochemically synthesized Fe_3O_4 nanoparticles, of which the median diameter was 16 nm, had a high crystallinity and a high saturation magnetization of 78 emu/g, and showed superparamagnetism. The wet mechanochemical process also contributed to the distribution of Fe_3O_4 nanoparticles in the HA matrix. The Fe_3O_4/HA nanocomposites were confirmed to have a good hyperthermia property through the measurement of temperature increase in an AC-magnetic field. For example, the 30 mass% Fe_3O_4/HA nanocomposites showed the temperature increase of 40°C after about 20 sec under a frequency of 600 kHz and an amplitude of 2.9 kA/m. Consequently, the Fe_3O_4/HA nanocomposites thus synthesized were found to be a promising material for hyperthermia therapy.

Acknowledgements

The author would like to thank Professor Kenya Murase of Osaka University for his support in measuring temperature profile of nanocomposites.

Author details

Tomohiro Iwasaki*

Department of Chemical Engineering, Osaka Prefecture University, Japan

References

[1] Murakami S, Hosono T, Jeyadevan B, Kamitakahara M, Ioku K. Hydrothermal synthesis of magnetite/hydroxyapatite composite material for hyperthermia therapy for bone cancer. Journal of the Ceramic Society of Japan 2008;116(9) 950–954.

[2] Andronescu E, Ficai M, Voicu G, Ficai D, Maganu M, Ficai A. Synthesis and characterization of collagen/hydroxyapatite: magnetite composite material for bone cancer treatment. Journal of Materials Science: Materials in Medicine 2010;21(7) 2237–2242.

[3] Inukai A, Sakamoto N, Aono H, Sakurai O, Shinozaki K, Suzuki H, Wakiya N. Synthesis and hyperthermia property of hydroxyapatite-ferrite hybrid particles by ultrasonic spray pyrolysis. Journal of Magnetism and Magnetic Materials 2011;323(7) 965–969.

[4] Tampieri A, D'Alessandro T, Sandri M, Sprio S, Landi E, Bertinetti L, Panseri S, Pepponi G, Goettlicher J, Bañobre-López M, Rivas J. Intrinsic magnetism and hyperthermia in bioactive Fe-doped hydroxyapatite. Acta Biomaterialia 2012;8(2) 843–851.

[5] Dong L, Zhu Z, Qiu Y, Zhao J. Removal of lead from aqueous solution by hydroxyapatite/magnetite composite adsorbent. Chemical Engineering Journal 2010;165(3) 827–834.

[6] Wang X. Preparation of magnetic hydroxyapatite and their use as recyclable adsorbent for phenol in wastewater. Clean—Soil, Air, Water 2011;39(1) 13–20.

[7] Xie H, Li X, Cheng C, Wu D, Zhang S, Jiao Z, Lan Y. Kinetic and thermodynamic sorption study of radiocobalt by magnetic hydroxyapatite nanoparticles. Journal of Radioanalytical and Nuclear Chemistry 2012;291(2) 637–647.

[8] Yang Z, Gong X, Zhang C. Recyclable Fe_3O_4/hydroxyapatite composite nanoparticles for photocatalytic applications. Chemical Engineering Journal 2010;165(1) 117–121.

[9] Liu Y, Zhong H, Li L, Zhang C. Temperature dependence of magnetic property and photocatalytic activity of Fe$_3$O$_4$/hydroxyapatite nanoparticles. Materials Research Bulletin 2010;45(12) 2036–2039.

[10] Covaliu CI, Georgescu G, Jitaru I, Neamtu J, Malaeru T, Oprea O, Patroi E. Synthesis and characterization of a hydroxyapatite coated magnetite for potential cancer treatment. Revista de Chimie 2009;60(12) 1254–1257.

[11] Iwasaki T, Sato N, Kosaka K, Watano S, Yanagida T, Kawai T. Direct transformation from goethite to magnetite nanoparticles by mechanochemical reduction. Journal of Alloys and Compounds 2011;509(4) L34–L37.

[12] Silva CC, Pinheiro AG, Miranda MAR, Góes JC, Sombra ASB. Structural properties of hydroxyapatite obtained by mechanosynthesis. Solid State Sciences 2003;5(4) 553–558.

[13] Shu C, Yanwei W, Hong L, Zhengzheng P, Kangde Y. Synthesis of carbonated hydroxyapatite nanofibers by mechanochemical methods. Ceramics International 2005;31(1) 135–138.

[14] Wu SC, Hsu HC, Wu YN, Ho WF. Hydroxyapatite synthesized from oyster shell powders by ball milling and heat treatment. Materials Characterization 2011;62(12) 1180–1187.

[15] Iwasaki T, Yabuuchi T, Nakagawa H, Watano S. Scale-up methodology for tumbling ball mill based on impact energy of grinding balls using discrete element analysis. Advanced Powder Technology 2010;21(6) 623–629.

[16] Iwasaki T, Sato N, Nakamura H, Watano S. Aqueous-phase synthesis of crystalline magnetite nanoparticles by a new mechanochemical method. In: proceedings of the 5th Asian Particle Technology Symposium, APT2012, 2–5 July 2012, Singapore.

[17] Lee J, Isobe T, Senna M. Magnetic properties of ultrafine magnetite particles and their slurries prepared via in-situ precipitation. Colloids and Surfaces A: Physicochemical and Engineering Aspects 1996;109, 121–127.

[18] Murase K, Oonoki J, Takata H, Song R, Angraini A, Ausanai P, Matsushita T. Simulation and experimental studies on magnetic hyperthermia with use of superparamagnetic iron oxide nanoparticles. Radiological Physics and Technology 2011;4(2) 194–202.

Chapter 8

Layers of Inhibitor Anion – Doped Polypyrrole for Corrosion Protection of Mild Steel

Le Minh Duc and Vu Quoc Trung

Additional information is available at the end of the chapter

1. Introduction

1.1. Theoretical background

Almost metals are in contact with wet atmosphere or another aggressive medium such as seawater. Therefore, the corrosion process always occurs on the metal surface. This is also a challenge for scientists to control and reduce the enormous damages due to corrosion. The term 'corrosion' refers to deterioration of materials due to the chemical reactions with the environment. Corrosion is involved in the conversion of the surface of metals in contact with corrosive medium into another insoluble compound. Corrosion is also defined as 'the undesirable deterioration' of a metal or an alloy i.e. an interaction of the metal with its environment affecting the main properties of the metal. Corrosion protection is required for a long life and economical use of equipment in technical processes [1, 2].

Corrosion preventing technology has many options, for instance cathodic protection, anodic protection, use of corrosion inhibitors, forming the precipitates on the metal surface and acting as passive layers, organic coatings etc. Among the methods to prevent corrosion of metals, protection by conducting polymers has been investigated extensively in the recent years [3-18]. This is considered as a possible alternative for friendly-environment coating because an electrochemical process could eliminate the use of toxic chemicals. There are many publications related to conducting polymer in corrosion protection. Conducting polymer can decrease the corrosion rate of many metals such as iron, mild steel, aluminium, magnesium and others [19-25].

Conducting polymer has been investigated extensively for corrosion protection of metal. It is observed that a conducting polymer film alone cannot protect an unnoble metal completely. With a galvanic coupling experiment it could be shown that the polymethylthiophene film did not act as a redox mediator, passivating the steel substrate within the defect and reoxidising itself by dissolved oxygen [26]. Polypyrrole could not provide anodic protection for iron [9, 27]. Conducting polymers like polyaniline, polypyrrole (PPy) etc. can improve the corrosion protection of unnoble metals but it is impossible that the porous films protect the metal surface completely. It is expected that protective properties of polypyrrole can be improved by dopant anion [28].

Counter anions, the so-called dopant anions play an important role in the development of physical properties and morphology. These actions of the anions could be:

- Electroneutralising: dopant anions neutralise the positive charges on the polymer backbone during synthesis of conducting polymers.

- Changing the morphology: the size of the dopant anion can control the microstructure and the porosity of the polymer film.

- Improving the conductivity: the interaction between the positive charges of polymer and anions can influence the conductivity of the polymer films.

- Stabilising the polymer films.

- Compatibility with other polymeric matrices.

- Corrosion inhibition: small dopant anions can be released from the polymer coating when the coating is reduced. If these anions have some inhibiting properties they can provide for some additional protection.

1.2. Mechanism for corrosion protection by polypyrrole

Corrosion at metal surfaces is a severe industrial problem. A large amount of metal is wasted by corrosion. It can cause tremendous economic damages. Minimising this corrosion can save substantial money and prevent accidents due to equipment failure. Corrosion has and continues to be the research object of scientists [1, 2].

Corrosion is an electrochemical process in nature. An anode (negative electrode), a cathode (positive electrode), an electrolyte (environment), and a circuit connecting the anode and the cathode are required for corrosion to occur. For simplicity, the corrosion process of iron in aqueous environment is discussed as a typical example.

The general reaction that occurs at the anode is the dissolution of metal atoms as ions:

$Fe = Fe^{2+} + 2e$

Electrons from the anode flow to the cathode area through the metallic circuit and force a cathodic reaction (or reactions) to occur. Depending on the pH of the electrolyte, different cathodic reactions can occur. In alkaline and neutral aerated solutions, the predominant cathodic reaction is

$O_2 + 2H_2O + 4e \rightarrow 4OH^-$

In aerated acids, the cathodic reaction could be

$O_2 + 4H^+ + 4e \rightarrow 2H_2O$

In deaerated acids, the cathodic reaction usually occurs is

$2H^+ + 2e \rightarrow H_2$

The corrosion product formed on iron surface in the presence of oxygen is:

$Fe^{2+} + 2OH^- \rightarrow Fe(OH)_2$

This hydrous ferrous oxide ($FeO.nH_2O$) or ferrous oxide $Fe(OH)_2$ composes a diffusion barrier layer on the surface. This layer is green to greenish black in colour. In the presence of oxygen Fe^{2+} is oxidised to Fe^{3+}. Ordinary rust is the product of this step. The formal reaction equation is

$4Fe(OH)_2 + O_2 + 2H_2O = 4Fe(OH)_3$

Hydrous ferric oxide is orange to red-brown in colour. It exists as nonmagnetic Fe_2O_3 (hematite) or as magnetic Fe_2O_3. $Fe_3O_4.nH_2O$ often forms a black intermediate layer between hydrous Fe_2O_3 and FeO. Hence, rust films can consist of up to three layers of iron oxides in different states of oxidation [29].

Beck et al. suggested a model of corrosion protection by PPy [25]. The initial fast corrosion was a superposition of cathodic film reduction and anodic oxidation. Cathodic process was the driving process. The second step was caused of nucleophilic molecules dissolved in the solid. Both processes were of pseudo-first-order.

Jude O. Iroh et al. suggested a corrosion protection mechanism of iron by PPy on the basis of EIS results [10]. The double bonds and the polar –NH group in the ring caused the strong adsorption of PPy and improved corrosion protection. PPy coating was acting as diffusion barrier and was inhibiting charge transfer.

Su and Iroh reported a large shift of the corrosion potential (E_{corr}) nearly 600 mV for PPy(oxalate) on steel compared to bare steel [10]. Reut et al. also recorded this shift of corrosion potential. The different shifts of E_{corr} could be explained by the different pretreatment of the substrate [30]. It was concluded from these results that PPy produced the significant ennobling of steel.

But controversial results were published by Krstajic et al. [9]. It was found that PPy(oxalate) did not provide anodic protection of mild steel in 0.1 M H_2SO_4. PPy was undoped in a short time. The dissolution of the steel continued in the pores of the coating. The mechanism of corrosion protection of steel by PPy is not yet fully understood and it is likely to change with the corrosion conditions [30].

PPy film doped with inhibitor anions such as molybdate were synthesised on mild steel in a one-step process. Corrosion tests indicated a significant improvement of the protective performance of PPy film. PPy coatings even prevented corrosion in defect of the coating. Now the corrosion protection mechanism of PPy with small defect on mild steel will be discussed.

A defect on PPy coating deposited on mild steel substrate may be produced either because of manufacturing or due to damage. When this occurs, bare mild steel is exposed to the corrosive atmosphere. The oxidation of substrate occurs. The following corrosion or anodic reaction takes place at the bottom of the defect:

$$Fe = Fe^{2+} + 2e \qquad (1)$$

The open circuit potential of the coated sample drops down to the corrosion potential of iron if the sample is immersed in NaCl (see Figure 11).

The electrons produced in (1) are consumed in cathodic reactions as followed:

$$O_2 + 2H_2O + 4e = 4OH^- \qquad (2)$$

$$PPy_{OX}(DOP) + ne = PPy_{RED} + DOP^{n-} \qquad (3)$$

Where PPy_{OX}, PPy_{RED} are the oxidised and reduced states of polypyrrole, respectively. DOP^{n-} is a dopant anion.

The PPy film is reduced (reaction 3) as a result of the galvanically coupling to metal substrate. Reaction (2) takes place within the defect as well as on PPy film if Ppy is conductive and can mediate the electron transfer [31, 32].

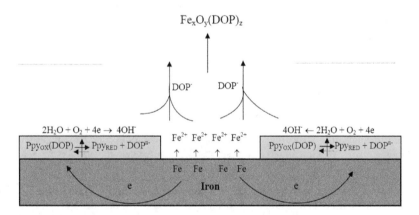

Figure 1. Model for the mechanism of corrosion protection of PPy proposed by Plieth et al. [34-36].

The dopant anions DOP^{n-} form insoluble salts or complexes $[Fe_xO_y(DOP)]$ with iron ions and can prevent further corrosion. In other words, the defect is repaired by the inhibitor anions produced during the reduction of PPy film.

Based on the model of Kinlen [33], the proposed mechanism of corrosion protection of PPy is developed as in figure 1 [34-36]. However, from the results of the OCP and EIS measurement re-oxidation of conducting polymer by of oxygen could not be found.

In most studies on the corrosion protection of mild steel by PPy, the role of the dopant anion as corrosion inhibitor has been investigated somewhere. The use of large anions such as polystyrensulfonate, dodecyl sulphate could improve the corrosion protection of the PPy film by preventing the penetration of chloride [20, 21]. Corrosion is not inhibited if the coating has small defects. The defect is protected only if inhibitor anions can diffuse to the defect. Thus, the mobility of dopant anions is one of the important parameters. The release of dopant anions from the polypyrrole film is the first step of corrosion inhibition. Further studies presented in this chapter are necessary to understand the role of dopants in PPy film for corrosion protection. In addition, the organic coatings (such as epoxy) containing nanocomposites based on suitable anion doped PPy seems to be an useful solution for application to replace the toxic cromate paitings. Therrefore, in this chapter the corrosion protection of mild steel of epoxy coatings using doped PPy nanocomposites is also presented.

2. Experimental and analytical methods

2.1. Chemicals

Pyrrole monomers (Aldrich, 98-99%), $LiClO_4$ (Fluka, P.A), $(C_4H_9)_4NBr$ (Merck, P.A), Na_2MoO_4 (Aldrich), Na_2MoO_4 (Acros) and NaCl (J.T.Baker, P.A). Clay obtained from Di Linh mine, Vietnam, was refined by suspension method and then was sodized. Epoxy resin was received from Dow (D.E.R 324); hardener DETA (Diethylentriamin) was also perchased from Dow (USA) with amount of 10%.

2.2. Equipment

The following equipment was employed:

* EG&G-263A Model potentiostat/galvanostat.

* IM6 and Zenium impedance measurement system of ZAHNER-Elektrik. The ZAHNER simulation software was integrated in this system. The frequency range used was 100 kHz – 0.1 Hz.

* The network analyser Advantest R3753BH was connected to the EG&G-263A Model potentiostat/galvanostat to measure the impedance of the quartz electrode in EQCM.

* SEM pictures were obtained with Zeiss DSM 982 Gemini microscope (Carl Zeiss, Germany).

* Raman Spectrometer Series 1000 – Renishaw.

2.3. Cell of measurements

2.3.1. Cell for electropolymerisation

The cell consisted of two parts which can be screwed together by Teflon screws. The working electrode (WE) was placed between of them. A silicon ring was used for sealing. The surface of the WE was 0.64 cm^2. The Pt sheet counter electrode (CE) was placed in a narrow slot. The distance between anode and cathode was 2.5 cm. The reference electrode was connected with the working electrode through a salt bridge.

2.3.2. Cell for EIS measurement

The EIS cell consisted of two parts which could be fixed to each other with four metal screws at the corners. The working electrode was placed between these parts. A silicon gasket defined the immersion surface of working electrode (0.125 cm^2). A glass container was placed on the upper part to hold the electrolyte. Pt net was the counter electrode.

2.4. Pretreatments of substrates

The substrates were pretreated as below:

- Mild steel (20 x 20 mm^2): Polishing with emery paper No 600; rinsing in ethanol in an ultrasonic bath for about 15 minutes; then drying in an N$_2$ stream.

- Passive film on mild steel (20 x 20 mm^2): Treating as described above; immersion in 0.1 M Na$_2$MoO$_4$ for 60 minutes, 30^{0}C; potentiostatic passivation at 0.5 V$_{SCE}$ for 1 hour; rinsing in distilled water and drying in an N$_2$ stream.

2.5. Condition for electrochemical polymerisation of polypyrrole

The PPy films were generated galvanostatically on the pretreated mild steel surface at a current density of 1.5 mA.cm^{-2} in an aqueous solution of 0.1 M pyrrole (from Aldrich 98%, keep at 4°C and distil in argon atmosphere before using) and 0.01 M sodium molybdate (pH=4.8). After forming, the sample was rinsed in distilled water and dried in nitrogen atmosphere. The film thickness was about 1–1.2 μm. For investigation of the release behaviour of molybdate anion during the reduction of PPy, the films were electrodeposited on Pt.

2.6. Condition for chemical preparation of molybdate anions doped polypyrrole/ montmorillonite nanocomposites PPy(MoO$_4$)/MMT

Nanomposites were prepared following a procedure described in Ref. [37, 38]. Firstly, a dispersion was prepared by mixing (for 30 min.) of 3.0 g Na$^+$-MMT, 0.87 g dopant and 1.0 mL pyrrole monomers in distilled water. Then 3.94 g (NH$_4$)$_2$S$_2$O$_8$ was added to the dispersion during stirring. The colour of the mixture was changed from grey to black. After 2 hours of stirring, the solid product was cleaned by distilled water, filtered and dried at 40-50^{0}C for 6 hours under low pressure.

2.7. Preparation of epoxy coatings

Epoxy coatings containing molybdate anions doped PPy/montmorillonite nanocomposites PPy(MoO$_4$)/MMT were prepared by spray. The thickness of the coatings are 50-60μm. The amounts of PPy(MoO$_4$)/MMT in the coatings are 2; 3; 4% (by weight).

2.8. Electrochemical impedance spectroscopy and its interpretation

Electrochemical Impedance Spectroscopy (EIS) primarily characterises a conducting polymer system in terms of electrical properties. The electrochemical behaviour of the polymer film is substituted by an equivalent circuit. The typical Bode plot and the equivalent circuit of a Ppy film are shown in Figure 2 and in Figure 3, respectively.

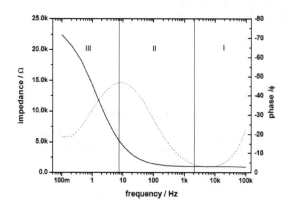

Figure 2. Bode of PPy film: Impedance (solid line) and phase angle (dot line)

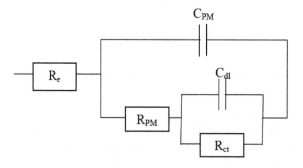

Figure 3. Equivalent circuit for fitting the impedance spectrum of PPy films [34-36].

The obtained impedance spectra of the polymer film can be divided in three regions:

- Region I (*high frequency > 1 kHz*): characterises the behaviour of the electrolyte. Phase angle is nearly zero (dotted line in Figure 2). The resistance of the electrolyte is described by Re in the equivalent circuit (Figure 3).

- Region II (*middle frequency*): shows the properties of the polymer film. The PPy film behaves as dielectricum with a capacitive impedance. The behaviour is simulated by a capacitance C_{PM} parallel to the resistance of the polymer film R_{PM}.

- Region III (*low frequency*): represents the interface polymer/substrate. C_{dl} and R_{ct} are the capacitance of double layer and the charge transfer resistance of the interface, respectively.

The experimental EIS data could be modelled by the equivalent circuit in Figure 3 using the fitting procedure of Zahner software.

3. Results and discussions

3.1. Molybdate anions doped polypyrrole films for corrosion protection

3.1.1. Electropolymerisation of pyrrole on mild steel

Electrochemical polymerisation of pyrrole on metals such as Fe, Zn, Al, and Mg is prevented by the oxidation of these metals because the oxidation potentials are lower than that of pyrrole. The dissolution of metals is so large that the PPy film has no adhesion to the substrate. This problem can be overcome by many methods [39], one of them is the proper selection of the electrolyte.

The role of molybdate as an inhibitor stored in the PPy film as well as the use of the PPy to protect mild steel from corrosion is discussed in the part of theoretical background.

3.1.2. Behaviour of mild steel in molybdate solution

Figure 4 shows the open circuit potential (OCP) vs. time curves of the mild steel electrode immersed in aerated and deaerated aqueous solution of 0.01 M Na_2MoO_4.

As seen in Figure 4, in both cases, the OCP shifts rapidly to positive potentials towards the passive region of mild steel. This potential rise can be ascribed to the reaction between molybdate anions and mild steel as soon as the electrode is immersed into the solution. The insoluble product can block the surface and make the surface potential more positive. This effect can be seen in deaerated medium. In the presence of oxygen, however, the OCP shift is faster and achieves larger positive potentials.

In other words, MoO_4^{2-} compound acts as an oxidant and passivates the surface of mild steel surface shortly even if without oxygen. Surface analysis of mild steel exposed to molybdate by XPS, AES and an electron microprobe confirms the presence of FeO.OH in combination with MoO_3 [40]. In contrast, the passivation of mild steel needs oxygen in solutions containing inhibitor such as oxalate, succinate, and phthalate [41].

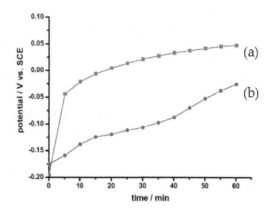

Figure 4. OCP – time curve of mild steel in aerated (a) and in deaerated (b) 0.01 M Na_2MoO_4

3.1.3. Electropolymerisation of pyrrole on mild steel

The PPy films were generated galvanostatically at a current density of 1.5 mA cm^{-2} in an aqueous solution of 0.1 M pyrrole and 0.01M sodium molybdate (pH=4.8). The total charge passed was 0.9 C cm^{-2}.

After formation, the sample was rinsed in distilled water and dried in nitrogen atmosphere. The potential-time curve for galvanic deposition of PPy(MoO$_4$) on mild steel is shown in Figure 5.

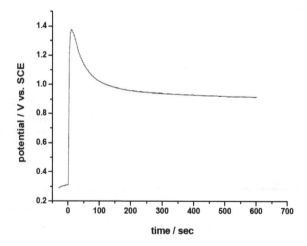

Figure 5. Electrochemical polymerization of pyrrole on mild steel (i = 1.5 mA.cm^{-2}; 0.01 M MoO$_4^{2-}$ pH = 4.8; 0.1 M pyrrole)

The polymerisation starts at a constant current of 1.5 mA.cm^{-2} for which the potential increases rapidly and then decreases slowly. After 100 seconds, the potential is stabilised at about 0.9 V$_{SCE}$ which corresponds to the oxidation potential of pyrrole.

This behaviour indicates that the dissolution of mild steel is prevented. Pyrrole can be oxidised on mild steel. However, the oxidation potential of pyrrole is higher than normal. The reasons may be: i) the barrier effect of the passivating layer of molybdate on the surface ii) the lower conductivity of the electrolyte.

It should be noted that the polymerisation in presence of molybdate on mild steel occurs without an induction period. This is in contrast to other procedures which have been reported [20, 42, 43]. The electropolymerisation of pyrrole in oxalate aqueous solution is a typical example to elucidate the role of molybdate in the polymerisation. An induction time is present during the polymerisation as shown in Figure 6.

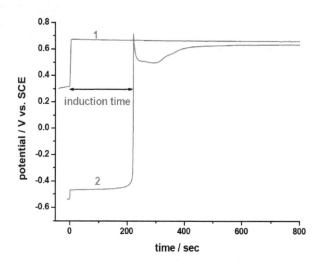

Figure 6. Potential – time curve for the electrodeposition of PPy on Pt(1) and on mild steel (2) from 0.1 M H$_2$C$_2$O$_4$; 0.1 M pyrrole, at 1 mA.cm^{-2}

Obviously, the surface of mild steel needs nearly 200 seconds to be passivated in oxalic acid (H$_2$C$_2$O$_4$). The induction time is attributed to the active dissolution of mild steel. This active range is assigned at the negative potential. Next, the formation of Fe-oxalate results in the potential shift towards positive potentials high enough for oxidation of pyrrole. Finally, the positive potential levels off until the end of polymerisation. The behaviour is different from that of the electropolymerisation on Pt. Pyrrole can be oxidised immediately on Pt after applying the current through the cell.

3.1.4. *Characterisation of PPy(MoO₄)/mild steel film*

3.1.4.1. *Film morphology*

Figure 7 shows SEM micrographs of PPy films doped with MoO_4^{2-} on mild steel (left) and on Pt (right). The total consumed charge for deposition was about 0.9 C cm^{-2}. The films formed on mild steel are homogenous, compact, but thinner than on Pt. The typical cauliflower structure of PPy films on mild steel is observed.

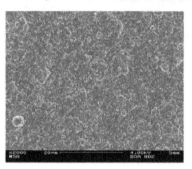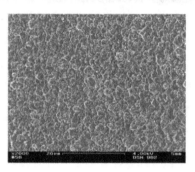

Figure 7. SEM pictures of PPy(MoO₄) on mild steel (left) and on Pt (right)

3.1.4.2. *Thickness of PPy films on mild steel*

A PPy film was formed on mild steel at the condition mentioned above, and then covered by conducting resin. A cross-section of the sample was made by cutting and polishing with 1200 emery paper. SEM and Energy Dispersive X-ray (EDX) measurements were carried out to determine the thickness of the PPy film.

As seen in Figure 8, three parts can be distinguished clearly. The Ppy film is determined in the middle part. The thickness of the Ppy film was estimated to about 1.2 μm, the charge consumed was 0.9 C cm^{-2}. The result is reproducible.

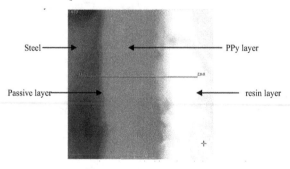

Figure 8. Figure 8. Cross-section SEM of PPy(MoO₄)/mild steel total charge passed is 0.9 C.cm^{-2}

It has been known that total charge for electropolymerisation of pyrrole is about 0.4 C cm$^{-2}\mu^{-1}$ on inert electrodes such as platinum. The value obtained in the current experiment is 0.75 C μ^{-1}cm^{-2}. The reason of this difference may be: a part of charge is used for the passivation of mild steel with molybdate.

3.1.4.3. XPS analysis

XPS spectra for molybdate in PPy(MoO$_4$) on mild steel presented in [44] showed the XPS surface analysis of the PPy(MoO$_4$) on mild steel. PPy was electrodeposited under similar conditions mentioned above.

The complex spectra of Mo 3d peaks correspond to the chemical states. The peaks are assigned: Mo1$_{3d5}$ (231 eV), Mo1$_{3d3}$ (233 eV), Mo2$_{3d5}$ (234.5 eV) and Mo2$_{3d3}$ (236 eV) [45]. The XPS spectra are supposed that molybdate should be in two compounds: [MoO$_4$]$^{2-}$ (62%) and [Mo$_7$O$_{24}$]$^{6-}$ (38%). Both types of molybdate are doped in PPy.

One should take into account the fact that molybdate exist in different forms depending on the pH of solution [46]. This relation can be presented:

$$[MoO_4]^{2-} \Leftrightarrow [Mo_7O_{24}]^{6-} \Leftrightarrow [Mo_8O_{26}]^{4-} \Leftrightarrow [Mo_{36}O_{112}]^{8-} \Leftrightarrow MoO_3 \cdot 2H_2O$$
$$\text{pH=6.5 – 4.5} \qquad \text{pH=2.9} \qquad \text{pH<2} \qquad \text{pH= 0.9}$$

In this condition of electropolymerisation (pH about 4.8), the presence of [MoO$_4$]$^{2-}$ and [Mo$_7$O$_{24}$]$^{6-}$ in polymer film is obvious.

3.1.5. Anion release from Ppy(MoO$_4$)/mild steel

Electrochemical behaviour of PPy on mild steel was characterised by EIS combined with cyclic voltammetry. Figure 9 shows the change of the resistance R$_{PM}$ and the capacitance C$_{PM}$ of PPy(MoO$_4$)/mild steel with the potential.

In the potential range from 0.6 V - 0.2 V$_{SCE}$, the PPy film is conductive; R$_{PM}$ is small (about 20 Ω). Following the negative scan, R$_{PM}$ increases gradually. The reduction of the PPy film begins at 0.2 V$_{SCE}$. The behaviour is reversible (see reverse scan) i.e. R$_{PM}$ decreases nearly to the original value. The PPy(MoO$_4$) film is now in the oxidised state.

The change of the film capacitance C$_{PM}$ is inversely proportional to that of R$_{PM}$. It decreases in the negative scan and increases in the positive scan gradually. It is observed, however, that the film capacitance does not return to the original value. The explanation for this phenomenon is that the transport of solvent (water) accompanies by the anion exchange. It leads to the conformation changes of the conducting polymer.

Hence, it can be concluded that:

- PPy(MoO$_4$) can be synthesised electrochemically on active metals like mild steel, in a one-step process. The dissolution of mild steel can be prevented with molybdate. The film is homogenous and adhesive on mild steel.

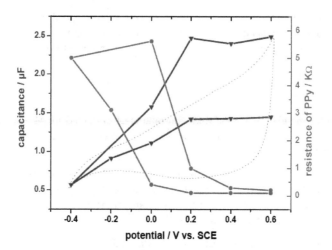

Figure 9. Change of R_{PM} and C_{PM} during reduction of PPy(MoO$_4$)/mild steel in 0.1 M (Bu)$_4$NBr, N$_2$, Dotted line is the CV

- PPy(MoO$_4$) film formed on mild steel has the same redox property as on Pt. During reduction in (Bu)$_4$NBr solution the decrease of film capacitance is observed. It means that molybdate releases from the film.

- There is no evidence of film decomposition. Corrosion test of PPy(MoO$_4$) on mild steel

3.1.6. Polarisation curves

The protective effect of PPy(MoO$_4$) film on mild steel was examined in aerated 0.1 M NaCl. The corrosion potential E_{corr} and corrosion current density i_{corr} was determined by extrapolation of anodic and cathodic Tafel lines.

Figure 10 indicates that E_{corr} shifts towards positive potentials (about 500 mV) and i_{corr} decreases about one order in magnitude as compared to the bare mild steel electrode. The polymer film can prevent the metal surface from corrosion. The inhibiting efficiency E of PPy(MoO$_4$) is obtained by equation:

$$E = \frac{i^{0} - i}{i^{0}}\%$$

Where i^{0}, i denote the corrosion current of bare mild steel and PPy(MoO$_4$)/mild steel, respectively. The inhibiting efficiency E is about 98% for PPy(MoO$_4$) on mild steel.

3.1.7. OCP measurement

The corrosion behaviour of mild steel covered by PPy(MoO$_4$) films was investigated by OCP-time measurements. The samples were immersed in a 0.1 M NaCl solution as corrosive medium and the OCP was recorded versus time. The protection time is characterised by the

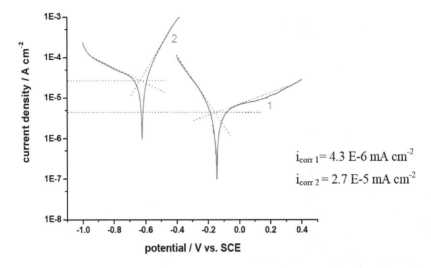

Figure 10. Potentiodynamic curve of PPy(MoO₄)/mild steel film (1) and bare mild steel (2) in 0.1 M NaCl, 1 mV.s⁻¹

time during which the OCP of the covered electrode remains in the passive state of mild steel before it drops down to the corrosion potential of unprotected mild steel.

The OCP-time curve of $PPy(MoO_4)$ coated on mild steel in NaCl is presented in Figure 11. The OCP is initially positive at about 0.35 V_{SCE} which corresponds to redox potential of PPy. The mild steel electrode maintains in its passive state for about 7 hours. Then, the potential sharply decreases to a second plateau at about -0.2 V_{SCE} and is stable there for about 5 hours. In this plateau, chloride anions have reached the metal surface through the pores of PPy film. The anodic reaction can take place and the polymer is reduced partially. This reduction causes the molybdate anion release which is needed to slow down the corrosion rate. This may be a reason why OCP is stable at plateau 2. The release of molybdate from the PPy film in this second plateau was confirmed by EIS measurement shown in Figure 12. The reduction of PPy film causes the increase of the film resistance (marked by an arrow) and the release of molybdate causes the decrease of the film capacitance.

Finally, OCP decreases towards the corrosion potential of the mild steel because of the large concentration of chloride in the polymer/mild steel interface, PPy cannot protect mild steel any longer.

This second plateau is only observed if the dopant anions of the conducting polymer have some ability to inhibit the corrosion reaction of mild steel [9, 20]. If the anions cannot give this protection, the second plateau is missing and the potential falls down to the corrosion potential of mild steel at the end of the first plateau [47].

$PPy(MoO_4)$ has shown the protective ability for mild steel. The corrosion potential is kept at the second plateau where mild steel is in a passive state.

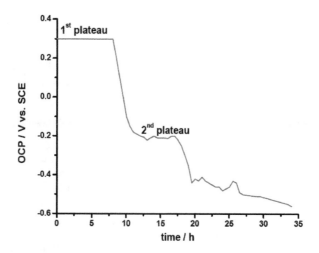

Figure 11. OCP-time curve of PPy(MoO$_4$)/mild steel (thickness of 1.5 μm) in 0.1 M NaCl

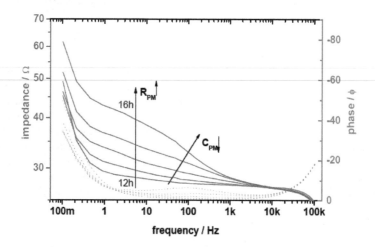

Figure 12. EIS spectra of the PPy(MoO4) film in the second plateau of the OCP in 0.1 M NaCl solution. Impedance (solid line) and phase angle (dotted line)

3.1.8. The role of molybdate passive layers in corrosion protection

It is clear that a passive layer of molybdate is formed on mild steel during the electropoly-merisation process. This layer can reduce the oxidation of mild steel and facilitates the polymerisation of pyrrole. In order to clarify the role of molybdate in this passive layer in corrosion protection, a PPy film with the non-inhibitive anions ClO$_4^-$ was electrodeposited on mild steel passivated with molybdate. The corrosion test was carried out in 0.1 M NaCl solution.

A mild steel electrode was passivated in molybdate solution with the following procedure: immersion in 0.1 M Na_2MoO_4 for 60 minutes, 30°C; potentiostatic passivation at 0.5 V_{SCE} for 1 hour; rinsing in distilled water and drying under N_2 stream [48]. The PPy(ClO_4) film was electrodeposited under these conditions: 0.1 M $LiClO_4$, 0.1 M pyrrole monomer, i = 1 mA cm^{-2}. The OCP of PPy(ClO_4) was recorded in 0.1 M NaCl.

Figure 13 shows the OCP-time curve of PPy(ClO_4)/passive layer/mild steel in 0.1 M NaCl. At the beginning, the OCP is still in the passive range of mild steel. PPy(ClO_4) can protect the substrate from corrosion. Nevertheless, this protection only remains for a short time (about 100 seconds). The penetration of chloride through the film is very fast and breaks down the passive film of molybdate formed in the pretreatment procedure. The OCP decreases sharply to the active potential range.

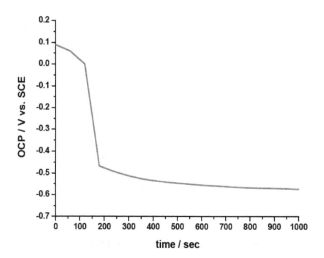

Figure 13. OCP-time vurve of PPy(ClO_4)/passive layer/mild steel in 0.1 M NaCl

This OCP measurement indicates that the passive film of molybdate on the mild steel cannot prevent the penetration of chloride and cannot reduce the corrosion rate. Molybdate in a passive layer under the polymer film does not play any role for corrosion protection.

3.1.9. The possibility of self-healing with PPy(MoO_4) film on mild steel

The self-healing action of PPy(MoO_4) film was investigated. On a fresh PPy(MoO_4)/mild steel film, a small defect (about 0.04 mm^2) was made with a needle. OCP–time curve was recorded in aerated 0.1 M NaCl as seen in Figure 14.

After immersion, the potential decreases immediately and then levels off at a potential of about -0.1 V for 4 hours. This observation could be explained as follows: the dissolution of mild steel at defect occurs immediately after the PPy film contacts with the corrosive environment. Because of the galvanical connection with mild steel, PPy will be reduced to compensate the redox process. This reduction is a driving-force for molybdate anions to be released from the PPy near the defect. A passive compound $Fe_xMo_yO_z$ is produced and it acts as inhibitor in the defect. It results in the maintenance of the potential for a certain time. The fluctuation of the OCP, numerous small spikes of potential are observed, is explained as the breakdown of the passive film by chloride and the re-passivation by MoO_4^{2-} in the defects. A small defect of the PPy film is protected from corrosion by the $PPy(MoO_4)$ film.

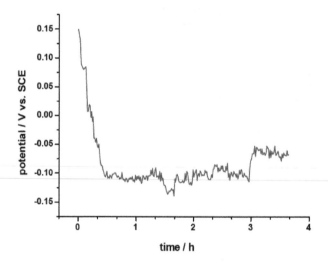

Figure 14. OCP of PPy(MoO₄)/mild steel with a defect (about 0.04 mm²) *in aerated 0.1 M NaCl*

The same experiment was carried out with a PPy film doped with PF_6^- as non-inhibitive anion. After passivating with molybdate, a mild steel electrode was covered with PPy in 0.1 M $(Bu)_4NPF_6$, 0.1 M pyrrole in dichloromethane at 1.5 mA cm⁻². A small defect was made with a needle (0.04 mm²) on the fresh film. As corrosive medium 0.1 M NaCl was also used.

The polarisation curves shown in Figure 15 are obtained on two PPy films with different dopant anions, namely PF_6^- (curve 1) and molybdate (curve 2). Although there is a molybdate passive layer, the corrosion potential of $PPy(PF_6)$ is still in the active range at the beginning of the experiment. No shift of the corrosion potential is observed here. This behaviour shows that the defect is attacked continuously by chloride. $PPy(PF_6)$ film cannot protect and repair this defect.

On the contrary, a positive shift of the corrosion potential and the decrease of the corrosion current are observed on $PPy(MoO_4)$/mild steel (curve 2). The defect is passivated for 4 hours.

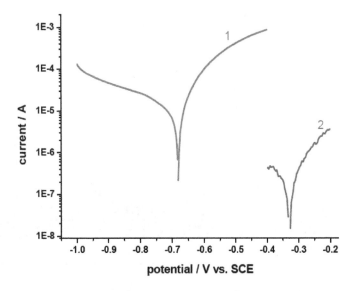

Figure 15. Fe/molybdate passive layer/PPy (PF$_6$), i=1.5 mA.cm^{-2}, in dichloromethane, about 1.5 µm; (2): Fe/PPy(MoO$_4$), ca. 1.5 µm, after dipping 4 hours

The corrosion potential remains in the passive range. The corrosion current for PPy(MoO$_4$) is nearly two orders of magnitude smaller than that of PPy(PF$_6$).

These results show that molybdate within a polymer plays an important role in self-healing of a defect. Passive layers containing molybdate on mild steel can reduce the dissolution of the active metal during the polymerisation but cannot act as corrosion inhibitor. The self-healing action for mild steel only takes place on PPy films doped with corrosion inhibitors such as molybdate.

3.1.10. Delamination

The corrosion process of PPy(MoO$_4$) on mild steel was investigated with the Scanning Kelvinprobe (SKP). The experimental set-up and the SKP measurements were made in MPI (Max-Planck Institute for Iron, Dusseldorf - Germany). The artificial defect was prepared on a part of PPy film shown in Figure 16.

The used SKP tip had a diameter of 100µm. Top-coat was polyacryl resin (BASF) applied on the film in order to avoid the difficulties resulting from the folding of the film during the delamination experiment. The humid atmosphere (93 – 95%) was controlled during the experiment.

PPy was formed on mild steel under the following conditions: 0.1 M pyrrole + 0.01 M molybdate aqueous electrolyte; current density 1.5 mA cm^{-2}. The SKP measurements were obtained in 0.1 M KCl solution. SKP experiment of PPy(MoO4)/mild steel in 0.1 M KCl presented in [34]

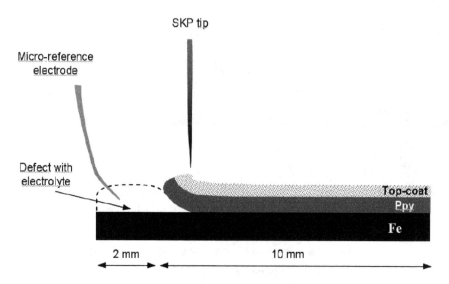

Figure 16. Preparation of iron electrode coated by PPy(MoO₄) folm for SKP experiment [34]

showed the profile of corrosion potential E_{corr} as function of the distance defect border for different times after contact of the defect with electrolyte. The features must be mentioned: (i) the region next to the defect where E_{corr} is similar corrosion potential of bare mild steel (-0.45 V_{SHE}). (ii) a region of abrupt increase low to high values of E_{corr}. This region shifts from left to right i.e. away from border with increasing time; (iii) a region where the adhesion is not yet lost [49].

It can be seen that the delamination of PPy film is very fast in KCl (about 1600 μm for 2 hours). The corrosion potential of the delaminated area remains in the active range potential of corroding iron (about 0.45 V_{SHE}). The passivation of the defect does not take place.

Because of their size, cations K^+ can move into the PPy film easier than the release of molybdate from the film during delamination process. The transport of anions in electrochemical experiment is only few micron while the delamination over hundreds micron, that leads to the predominately cation incorporation into the film for charge compensation. The amount of molybdate is not enough to passivate the large defect.

The delamination of PPy(MoO₄) film on mild steel investigated further in 0.1 M (Bu)₄NCl solution is shown in Figure 17. The delamination is much slower than that in small cation solution. Now the delamination front reaches 1600μm after 33 hours instead of 2 hours in KCl solution. The reduction of PPy at the defect is slowed down. Obviously, the size of cation in electrolyte had effects on the delamination process. The incorporation of cation $N(Bu)_4^+$ is hindered because of their size. The release of molybdate anions is predominant.

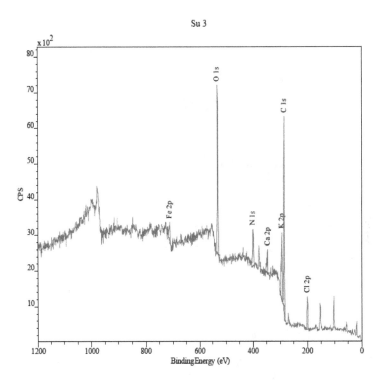

Figure 17. XPS measurement of PPy(MoO₄) on mild steel after delamination in 0.1 M (Bu)₄NCl [44]

The release of molybdate in delamination process is confirmed by the XPS experiment of PPy(MoO₄)/mild steel in 0.1 M (Bu)₄NCl [34]. After delamination, the PPy film was peeled off and molybdate was detected by XPS in the film. No signals of molybdate were found, only iron was seen in the spectra. The presence of iron may come from the corrosion process. The molybdate has moved to the defect during the corrosion process. The same SKP experiment was repeated and the amount of molybdate within the PPy film was measured by XPS at the defect and at the interface polymer/metal. The result is presented in Figure 18.

It can be seen that there is a difference of the amount of molybdate at the defect and at the interface polymer/substrate. At the defect, molybdate anions are expelled from the polymer due to the reduction of PPy when the defect is connected with the substrate galvanically. The amount of molybdate is consumed to form the passive layer and to suppress the cathodic delamination. Nevertheless, the PPy film is still in the oxidised state in the undelaminated region of PPy (point 2, Figure 18). Molybdate anions still remain in the PPy film. This is the reason why the quantity of molybdate in the PPy film at the defect is smaller than at the interface where the PPy protects the metal.

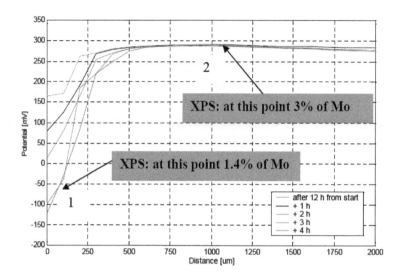

Figure 18. Quality of molybdate in the PPy film at the defect (1) and at the interface polymer/substrate (2) after delamination in 0.1 M (Bu)₄NCl [44]

3.1.11. Raman spectroscopy of PPy(MoO₄)

During the corrosion process, PPy will be reduced because a galvanic cell is established between PPy and mild steel. Raman spectroscopy was used to characterise the state of PPy, oxidised or reduced.

PPy(MoO₄)/mild steel was prepared in aqueous solution of 0.1 M pyrrole + 0.01 M molybdate at 1.5 mA cm⁻². Three states of PPy (fully oxidised, partially reduced and totally reduced) were determined through the OCP obtained by dipping samples in 0.1 M NaCl. Raman spectra of these samples are shown in Figure 19.

Several bands are representative for the oxidised state. The band 1600 cm⁻¹ belongs to the inter-ring (C=C) of oxidized PPy (0.1 V). It shifts towards low wavenumbers in more negative potential (1597 cm⁻¹ at –0.4 V and 1594 cm⁻¹ at –0.6 V). The bands 1052 cm⁻¹, 1083 cm⁻¹ are assigned to the C-H in plane deformation [50, 51]. They are also shifted to lower wavenumbers during the corrosion process. The Raman peaks of the dopant anion shift from 931 cm⁻¹ to 920 cm⁻¹ in the reduced state [52]. The PPy(MoO₄)/mild steel film is in the oxidised state while it protects mild steel substrate and is progressively changed to the reduced state during the protection progress.

The properties of PPy(MoO₄) can be summarised:

- PPy(MoO₄) films covered on mild steel have the effect of corrosion inhibition. The E_{corr} is in the passive range of potential in chloride solution. The potential shift is nearly 500 mV$_{SCE}$ compared to bare mild steel. At the same time, i_{corr} decreases one order of magnitude when

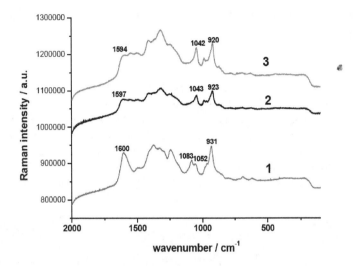

Figure 19. Raman spectra of PPy in different states: (1): complete oxidation (after formation), (2): partial reduction and (3): complete reduction

$PPy(MoO_4)$ covers on mild steel. The protective efficiency is fairly high (about 98%), indicating a good property of $PPy(MoO_4)$ in corrosion protection for mild steel.

- The self-healing effect of $PPy(MoO_4)$ film can be observed by OCP measurement. In contrast, $PPy(PF_6)$ cannot prevent corrosion of mild steel even with no defect. Molybdate in corrosion protection of mild steel is acting as anodic inhibitor.

- The release of the molybdate is necessary for self-healing of defect by PPy. The reduction of the PPy film during the corrosion process is the driving-force to release inhibitor anions. Raman experiments show that PPy changes from the oxidised state to the reduced state during the corrosion process. This observation is reported elsewhere [50, 53]. XPS results confirm this observation.

- The EQCM and EIS results in aqueous solution indicate that the reduction of $PPy(MoO_4)$ is accompanied by a mixed anion/cation transport and by water uptake. However, it is possible to show that the medium size molybdate anion can be released from the film to improve the protective properties of PPy film deposited on mild steel.

- The Scanning Kelvinprobe experiments of $PPy(MoO_4)$ film show that the release of molybdate can be found when the film is reduced in a large size cation solution. The delamination can be stopped. On the contrary, the delamination of $PPy(MoO_4)$ film is fast in a small cation solution. The incorporation of cations is predominant. Therefore, the delamination cannot be hindered [34].

- The release behaviour of molybdate from the PPy film depends much on the size and mobility of cations in the electrolyte.

3.2. Epoxy coatings containing molybdate anions doped polypyrrole/montmorillonite nanocomposites

3.2.1. SEM images

Figure 20 shows SEM micrographs of PPy (a), MMT (b) and PPy(MoO$_4$)/MMT nanocomposite (c). As shown in Figure 20, PPy is black powder (Figure 20a) and MMT consists of platelet particles accumulating each other and as crystals (Figure 20b). Compared with MMT (Figure 20b), the hackly surface of PPy(MoO$_4$)/MMT nanocomposite can be seen due to the deposition of the PPy onto the layer surface of MMT (Figure 20c). At high magnification, it is easier to see the flaky structure of MMT (Figure 20c). One can observe (as shown in Figure 20c at the tip of the arrow) the antenna-like PPy "stretching out" from the layer surface of MMT.

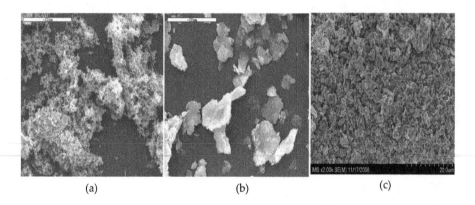

(a)   (b)   (c)

Figure 20. SEM images of PPy (a), MMT (b) and PPy/MMT nanocomposite (c)

3.2.2. XRD patterns

The XRD patterns of the materials before and after polymerization are shown in Figure 21. At first, Na$^+$-MMT was mechanically stirred for 30 min as reference. However, there are no changes in the XRD patterns of MMT before and after stirring. Therefore, the stirring does not affect the crystallinity of the MMT itself. The diffraction peak of Na$^+$-MMT was observed at $2\theta = 7.0°$, therefore, the basal spacing of Na$^+$-MMT was 1.22 nm (Fig. 21a). The intercalation of pyrrole monomers and dopant into MMT is shown in Figure 21b. The basal spacing increased from 1.22 nm to 1.58 nm ($2\theta = 5.8°$), indicating the expansion of the interlayer space (d-expansion) by 0.36 nm; and the successful intercalation by the mechanical intercalation method. The diffraction peaks of the products after polymerization were shifted to a higher angle than those before polymerization as shown in Figure 21c, indicating the synthesis of PPy in the clay layers. As a result, the the basal spacing of monomer-absorbed MMT changed from 1.58 nm to 1.42 nm ($2\theta = 6.0°$). They are in the agreement with other publications [54].

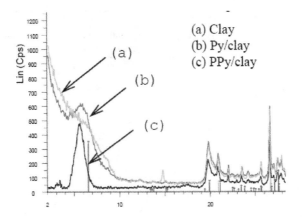

Figure 21. XRD patterns of MMT, monomer-absorbed MMT and PPy/MMT nanocomposites

3.2.3. Raman spectra

Figure 22 presents the Raman spectrum of PPy/MMT nanocomposite measured at 514 nm with 1 mW laser power. Table 1 gives the assignments of some Raman bands and compares the frequencies of the various vibration modes collected on the PPy/Ag spectra and with those theoretically calculated by Faulques *et al.* [55]. According to the theoretical values reported by Faulques et al. [55], the vibration frequency of the C=C double-bond of the PPy in the oxidized form (at 1593 cm^{-1}) is greater than that in the reduced form (Table 1). This evedence shows that molipdate anion was doped into PPy.

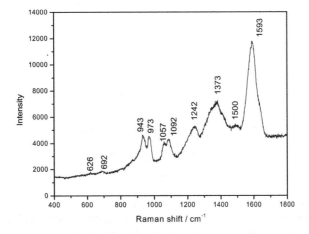

Figure 22. Raman spectrum of PPy(MoO$_4$)/MMT nanocomposite

Wavenumber (cm⁻¹)			
Theoretical calculation [55] (oxidized PPy)	PPy on Ag [56] (oxidized PPy)	PPy(MoO₄)/MMT (oxidized PPy)	Assignment
1676.6	1584	1593	C=C ring stretching
1524.4	1414	1500	C-N stretching
1307.2	1327	1373	
-	1258	1242	
1049.4	1046	1057; 1092	C-H in plane deformation
955.8	989	973	C-H stretching
-	938	943	

Table 1. PPy/MMT nanocomposites and their components

3.2.4. EDX spectra

In order to determine the presence of molipdate anion in the synthetic nanocomposites EDX spectra were used. The EDX result of C6 is presented in Figure 23. The amount of element Mo in the nanocomposite PPy/MMT nanocomposite is 5,55%.

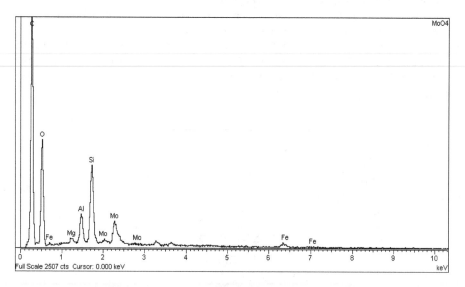

Figure 23. EDX spectra of PPy(MoO₄)/MMT nanocomposites

As seen in Figure 23 and Table 2, the amont of element Mo is 2.93% weight. It means that Na₂MoO₄ occupied 6.29% weight. In addition, the presence of other elements Si, Al, Mg, Fe, O in MMT and C in PPy component were presented in Fugure 23. Other peaks corresponding to hydrogen and nitrogen did not disappeared in EDX spectra.

Element	% Weight	% Element
C	54,53	64,15
O	37,07	32,74
Mg	0,27	0,16
Al	1,39	0,73
Si	3,33	1,68
Fe	0,46	0,12
Mo	2,93	0,43

Table 2. Amount of elements in PPy(MoO$_4$)/MMT nanocomposites

3.2.5. Thermal analyses

Thermal analyses of PPy(MoO$_4$)/MMT nanocomposites have been carried out. Figure 24 shows the thermal analyses curves of PPy(MoO$_4$)/MMT nanocomposites. Under 120^{0}C, the weight reduce origined from water inside samples. The strong reduce in this temperature range can be explained by the hydrophilic property of MMT and the oxidized state of PPy. It is also the source of the wide band between 4000 and 2500 cm^{-1} in the IR spectra. In the range of 120-330^{0}C, the weight reduces are very small, corresponding to the decomposition of redundant monomers, oligomers. At higher temparature (300-700^{0}C), the change of weight is attributed to the decomposition of the oxidized PPy. From Figure 24, the amounts of PPy in the PPy(MoO$_4$)/MMT nanocomposites are approximately calculated to be 16%.

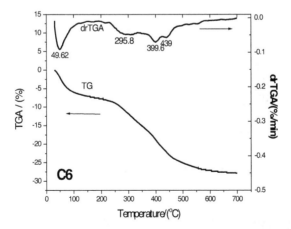

Figure 24. TGA curves and PPy(MoO$_4$)/MMT nanocomposite

3.2.6. Tests of corrosion protection

Figure 25 shows the OCP-time curves of steels covered by epoxy coatings contaning PPy(MoO$_4$)/MMT nanocomposite. It shows that the OPC of steel covered by epoxy was corroded after 2 weeks of immersion in NaCl solution. The OCP of the steel covered by epoxy coatings PPy(MoO$_4$)/MMT nanocomposite shows more possitive. Firstly the OCP of these samples gradually decreased and then increased (Figure 25). The reduction of OCP could be explained by the penetration of the corrosion medium. These results are in good agreement with that preasented in Figure 12. However, the OCP of these samples then incresed. These results shows the same phenomena of the sample PPy(MoO$_4$)/mild steel presented in Figure 11. When epoxy coating containing 2% PPy(MoO$_4$)/MMT, the OCP moved to anodic region from 7th day to 21st day and then keft plateau untill 42nd day. After that the OCP of A3 reduced to the value of -0.389VSCE on 56th day. In the case of epoxy containing 3% PPy(MoO$_4$)/MMT (sample A3), the OCP moved to anodic region from 7th to 28th day and keft plateu untill 42th day. On the 56th day, the OCP approached the value of -0.296V$_{SCE}$. When epoxy coating containing 4% PPy(MoO$_4$)/MMT, the OCP of the sample keft the constant value in the first week showing that the barier role of coating containing MMT. Then the OCP moved ossitively from 21st day to 28th day. This region could be explained by the by the penetration of the corrosion medium as weel as the situation of samples A2 and A3. After 42nd day, the OCP of sample A1 increased to -0.0087VSCE and keft plateau. This reason could be explained by the role of molybdate inhibitor anions. These rerults are in good agreement with that of PPy(MoO$_4$)/mild steel presented in section 3.1.

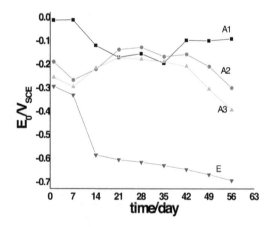

(A1: Epoxy containing 4% PPy(MoO$_4$)/MMT ; A2: Epoxy containing 3% PPy(MoO$_4$)/MMT ; A3: Epoxy containing 2% PPy(MoO$_4$)/MMT ; E: epoxy)

Figure 25. OCP-time curves of steel samples coverd epoxy coatings

In comparision with epoxy coating (sample E), all three epoxy coatings containing $PPy(MoO_4)/$ MMT could prevent mild steel from corrosion (Figure 25).

Figure 26 shows the impedance – time curves at 1Hz of steels covered by epoxy coatings containing. At low frequency (1 Hz) the impedance of the system equals to the value of resistance of epoxy coating (Figure 3). Generally, the imdepance of epoxy coatings gradually reduced during time of immersion. These results could be explained by the penetration of the corrosion medium. Epoxy coating had lowest impedance because of no nanocomposites inside. The samples of A2 and A3 had the impedances of around 10^6 Ω and it was stable throughout experiments. The sample A4 epoxy coating containing 4% $PPy(MoO_4)/MMT$ had highest impedance value at 1 Hz. It even increased after the 56th day (Figure 26). This reason could be explained by the role of molydate dopants.

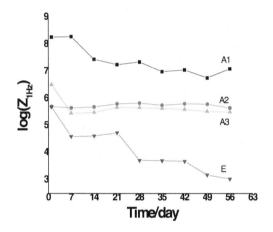

(A1: Epoxy containing 4% $PPy(MoO_4)/MMT$; A2: Epoxy containing 3% $PPy(MoO_4)/MMT$; A3: Epoxy containing 2% $PPy(MoO_4)/MMT$; E: epoxy)

Figure 26. Impedance – time curves of steel samples coverd epoxy coatings

4. Conclusion

Inhibitors used as dopant anions in polymer films are responsible for the anticorrosion behaviour of PPy. The PPy film can work as self-repairing for small defects of the film. Molybdate was built-in into the film as dopant anion. The results of XPS revealed that molybdate exist in two types: $[MoO_4]^{2-}$ (62%) and $[Mo_7O_{24}]^{6-}$ (28%). The film was conductive, homogenous, and compact. Cyclic voltammograms have shown that the film was active. By scanning the potential, the Ppy film changed from the oxidised to the reduced state and at the same time the anions were released from the polymer. To support this observation, Electrochemical Impedance Spectroscopy (EIS), Electrochemical Quartz Crystal Microbalance (EQCM) and Raman Spectroscopy were used. EIS has indicated the change of the resistance R_{PM} and the capacitance C_{PM} of the PPy film during reduction. EQCM has shown: the mass of the Ppy film decreased in the cathodic region and increased in the anodic region. The anion flux was also observed in Scanning Kelvinprobe (SKP) and X-ray Photoelectron Spectroscopy (XPS) experiments. The release of anion is one of the important factors of the corrosion protection property of the PPy film. However, the release behaviour of molybdate anions depends much on the size of cations in the electrolyte. At negative potentials, the incorporation of cations is predominant. Probably, molybdate is a medium size anion and not mobile enough to compete with the cations in the electrolyte for the moving in or out of the PPy film.

It should be also noted that $PPy(MoO_4)$ can be electrodeposited on mild steel in a one-step process. No induction period was observed during polymerisation. $PPy(MoO_4)$ shifted the corrosion potential of mild steel into the passive range. The corrosion current decreased simultaneously. The role of molybdate in corrosion protection was also investigated for Ppy films with small defects. The Open Circuit Potential (OCP) showed the fluctuations around -0.2 V_{SCE}. It means that the defect was passivated / repassivated continuously for 4 hours. It was assumed that molybdate was released from Ppy, move to the defect and act as corrosion inhibitor by forming a complex with iron ion.

The application of $PPy(MoO_4)$/MMT nanocomposites in corrosion protection for mild steel was also investigated in this work. Mixture of core-shell particles with a polymer was used as primer coatings. The positive effect on the corrosion protection of the coating was illustrated by the increase of the coating resistance and the stabilisation of the film capacitance during immersion in corrosive medium. These rerults show the promising application potential to exchange the paints containing toxic cromate.

Author details

Le Minh Duc[1] and Vu Quoc Trung[2*]

*Address all correspondence to: vuquoctrungvn@netnam.vn

1 Faculty of Chemical Engineering, Danang University of Technology, Danang, Vietnam

2 Faculty of Chemistry, Hanoi National University of Education, Hanoi, Vietnam

References

[1] C. M. A. Brett, A. M. O. Brett, Electrochemistry: Principles, Methods and Application, 1993.

[2]   R. W. Revie, Uhlig's Corrosion Handbook, John Wiley & Sons Inc, 2000.

[3]   G.G. Wallace, G.M. Spinks, L.A.P. Kane-Maguire, P.R. Teasdale, Conductive Electro-active Polymers: Intelligent Materials Systems, CRC press, New York, 2003.

[4]   W.K. Lu, R.L. Elsenbaumer, in: L. Rupprecht (Ed.), Conducting Polymers and Plastics in Industrial Applications, SPE, Norwich, NY, 1999, 195.

[5]   N. Ahmad, A.G. MacDiarmid, Synth. Met., 78 (1996), 103.

[6]   T. Schauer, A. Joos, L. Dulog, C.D. Eisenbach, Prog. Org. Coat., 33 (1998), 20.

[7]   R.N. Rothon, Adv. Polym. Sci., 139 (1999), 67.

[8]   A.M. Thayer, Chem. Eng. News, 81 (35) (2003), 15.

[9]   N. V. Krstajic, B. N. Grgur, S. M. Jovanovic, V. Vojnovic, Electrochimica Acta, 42 (1997), 1685.

[10] Jude O. Iroh, W. Su, Electrochimica Acta, 46 (2000), 15.

[11] Michael Rohwerder, Adam Michalik, Electrochimica Acta, 53 (2007) 1300–1313

[12] G. Mengoli, M.T. Munari, P. Bianco, M.M. Musiani, J. Appl. Polym. Sci., 26 (12) (1981), 4247.

[13] D.W. Deberry, J. Electrochem. Soc., 132 (5) (1985), 1022.

[14] B. Wessling, Mater. Corros., 47 (1996), 439.

[15] J. Reut, A. O° pik, K. Idla, Synth. Met., 102 (1999), 1392.

[16] T.D. Nguyen, M. Keddam, H. Takenouti, Electrochem. Solid-State Lett., 6 (2003), B25.

[17] M. Rohwerdera, Le Minh Duc, A. Michalika, Electrochimica Acta, 54 (2009), 6075.

[18] G. Williams, H.N. McMurray, Electrochem. Solid State Lett., 8 (2005), B42.

[19] Victoria Johnston Gelling, Michelle M. Wiest, Dennis E. Tallman, Gordon P. Bierwagen, Gordon G. Wallace, Prog. in Org. Coatings, Volume, 43 (1-3) (2001), 149.

[20] H. N. T. Le, B. Garcia, C. Deslouis, Q. L. Xuan, J. of App. Electrochem., 32 (2002) 105.

[21] N. T. H. Le, B. Garcia, A.Pailleret, C. Deslouis, Electrochimica Acta 50, (2005), 1747.

[22] A. Michalik, M. Rohwerder, Z. Phys. Chem., 219 (2005), 1547.

[23] G. Paliwoda-Porebska, M. Stratmann, M. Rohwerder, U. Rammelt, L. Minh Duc, W. Plieth, J. Solid State Electrochem., 10 (2006), 730.

[24] G. Paliwoda, M. Stratmann, M. Rohwerder, K. Potje-Kamloth, Y. Lu, A.Z. Pich, H.-J. Adler, Corros. Sci., 47 (2005), 3216.

[25] F. Beck, U. Barsch, R. Michaelis, J. of Electroanalytical Chemistry, 351 (1993), 169.

[26] U. Rammelt, P. T. Nguyen, W. Plieth, Electrochimica Acta, 48 (2003), 1257.

[27] U. Rammelt, P. T. Nguyen, W. Plieth, Electrochimica Acta, 46 (2001), 4251.

[28] M. Kendig, M. Hon, L. Warren, Progress in Organic Coating, 47 (2003), 183.

[29] Pierre R. Roberge, Handbook of corrosion engineering, McGraw-Hill, 2000.

[30] G. M. Spinks, A. J. Dominis, G. G. Wallace, D. E. Tallman, J. of Solid State Electro-chem., 6 (2002), 85.

[31] P. J. Kinlen, D. C. Silverman, C. R. Jeffreys, Synthetic Metals, 85 (1997), 1327.

[32] J. He, V. J. Gelling, D. E. Tallman, G. P. Bierwagen, G. G. Wallace, J. of Electrochem. Soc., 147 (2000), 3667.

[33] P. J. Kinlen, V. Menon, Y. Ding, J. of Electrochem. Soc., 146 (1999), 3690.

[34] G. Paliwoda-Porebska, M. Rohwerder, M. Stratmann, U. Rammelt, L. M. Duc, W. Plieth. J. of Solid State of Electrochem., 10 (2006), 730.

[35] U. Rammelt, L. M. Duc, W. Plieth, J. of Appl. Electrochem., 35 (2005),1225.

[36] W. Plieth, A. Bund, U. Rammelt, S. Neudeck, L.M.Duc, Electrochimica Acta, 2005.

[37] Yongqin Han, Polymer composite, 30 (1) (2009), 66.

[38] S.H. Hong, B.H. Kim, J. Joo, J.W. Kim, Hyung J. Choi, Current Applied Physics, 1 (6), (2001), 447.

[39] S. U. Rahman, M. S. Ba-Shamakh, Synthtic Metals, 140 (2004), 207.

[40] V. S. Sastri, Corrosion inhibitor- Principles and Applications, Wiley, 1998.

[41] G. Reinhard, M. Radtke, U. Rammelt, Corrosion Science, 33 (1992), 307.

[42] F. Beck, R. Michaelis, F. Schloten, B. Zinger, Electrochimica Acta, 39 (1994), 229.

[43] W. Su, Jude O. Iroh, Electrochimica Acta, 44 (1999), 2173.

[44] G.Paliwoda-Porebska, PhDThesis, Bochum Universtity, Germany, 2005.

[45] E. Almeida, T. C. Diamantino, M. O. Figueiredo, C. Sa, Surface and Coatings Tech., 106 (1998), 8.

[46] T. V. Vernitskaya, O. N. Efimov, A. B. Gavrilov, Russ. J. of Electrochem., 30 (1994), 1022.

[47] M.C. Bernard, S. Joiret, A. Hugo-Le Goff, P.V. Phong, J. of Electrochem. Soc., 148 (2001), B12.

[48] K. Aramaki, Corrosion science, 42 (2000), 1975.

[49] A. Leng, H. Streckel, M. Stratmann, Corrosion Science, 41 (1999), 547.

[50] H. N. T. Le, M. C. Bernard, B. G. Renaud, C. Deslouis, Synthetic Metals, 140 (2004), 287.

[51] Y. Chuan Liu, B. J. Hwang, Synthetic Metals, 113 (2000), 203.

[52] F. Chen, G. Shi, M. Fu, L. Qu, X. Hong, Synthetic Metals, 132 (2003), 125.

[53] T. D. Nguyen, T. Anh Nguyen, M. C. Pham, Electroanalytical Chemistry, 572 (2004), 225.

[54] S. Yoshimoto, F. Ohashi, Y. Ohnishi and T. Nonami, Chem. Commun., 17, (2004), 1924.

[55] A. Faulques, W. Wallnoefer, H. Kuzmany, J. Chem. Phys., 90 (12), (1989), 7585.

[56] M. Bazzaoui, E. A. Bazzaoui, L. Martins and J. I. Martins, Synthetic Metals, 128 (1), (2002), 103.

Permissions

The contributors of this book come from diverse backgrounds, making this book a truly international effort. This book will bring forth new frontiers with its revolutionizing research information and detailed analysis of the nascent developments around the world.

We would like to thank Prof. Yitzhak Mastai, for lending his expertise to make the book truly unique. He has played a crucial role in the development of this book. Without his invaluable contribution this book wouldn't have been possible. He has made vital efforts to compile up to date information on the varied aspects of this subject to make this book a valuable addition to the collection of many professionals and students.

This book was conceptualized with the vision of imparting up-to-date information and advanced data in this field. To ensure the same, a matchless editorial board was set up. Every individual on the board went through rigorous rounds of assessment to prove their worth. After which they invested a large part of their time researching and compiling the most relevant data for our readers. Conferences and sessions were held from time to time between the editorial board and the contributing authors to present the data in the most comprehensible form. The editorial team has worked tirelessly to provide valuable and valid information to help people across the globe.

Every chapter published in this book has been scrutinized by our experts. Their significance has been extensively debated. The topics covered herein carry significant findings which will fuel the growth of the discipline. They may even be implemented as practical applications or may be referred to as a beginning point for another development. Chapters in this book were first published by InTech; hereby published with permission under the Creative Commons Attribution License or equivalent.

The editorial board has been involved in producing this book since its inception. They have spent rigorous hours researching and exploring the diverse topics which have resulted in the successful publishing of this book. They have passed on their knowledge of decades through this book. To expedite this challenging task, the publisher supported the team at every step. A small team of assistant editors was also appointed to further simplify the editing procedure and attain best results for the readers.

Our editorial team has been hand-picked from every corner of the world. Their multi-ethnicity adds dynamic inputs to the discussions which result in innovative outcomes. These outcomes are then further discussed with the researchers and contributors who give their valuable feedback and opinion regarding the same. The feedback is then collaborated with the researches and they are edited in a comprehensive manner to aid the understanding of the subject.

Apart from the editorial board, the designing team has also invested a significant amount of their time in understanding the subject and creating the most relevant covers. They scrutinized every image to scout for the most suitable representation of the subject and create an appropriate cover for the book.

The publishing team has been involved in this book since its early stages. They were actively engaged in every process, be it collecting the data, connecting with the contributors or procuring relevant information. The team has been an ardent support to the editorial, designing and production team. Their endless efforts to recruit the best for this project, has resulted in the accomplishment of this book. They are a veteran in the field of academics and their pool of knowledge is as vast as their experience in printing. Their expertise and guidance has proved useful at every step. Their uncompromising quality standards have made this book an exceptional effort. Their encouragement from time to time has been an inspiration for everyone.

The publisher and the editorial board hope that this book will prove to be a valuable piece of knowledge for researchers, students, practitioners and scholars across the globe.

List of Contributors

Jairo Tronto
Universidade Federal de Viçosa - Instituto de Ciências Exatas e Tecnológicas - Campus de Rio Paranaíba - Rio Paranaíba - MG, Brazil

Ana Cláudia Bordonal and João Barros Valim
Universidade de São Paulo - Faculdade de Filosofia Ciências e Letras de Ribeirão Preto - Departamento de Química - Ribeirão Preto – SP, Brazil

Zeki Naal
Universidade de São Paulo - Faculdade de Ciências Farmacêuticas de Ribeirão Preto – Departamento de Física e Química - Ribeirão Preto – SP, Brazil

Hagay Moshe and Yitzhak Mastai
Department of Chemistry and the Institute of Nanotechnology, Bar Ilan University, Israel

Alex Lugovskoy and Michael Zinigrad
Chemical Engineering Department, Ariel University Center of Samaria, Ariel, Israel

Hiromi Shima and Soichiro Okamura
Department of Applied Physics, Tokyo University of Science, 1-3, kagurazaka, Shinjukuku, Tokyo, Japan

Hiroshi Naganuma
Department of Applied Physics, Graduate school of Engineering, Tohoku University, Japan

V. P. Zlomanov
Department of Chemistry, Moscow State University, Moscow, Russia

A.M. Khoviv and A.Ju. Zavrazhnov
Department of Chemistry, Voronezh State University, Voronezh, Russia

Rathinam Chandramohan
Department of Physics, Sree Sevugan Annamalai College,Devakottai, Tamil Nadu, India

Jagannathan Thirumalai and Thirukonda Anandhamoorthy Vijayan
Department of Physics, B.S. Abdur Rahman University, Vandalur, Chennai, Tamil Nadu, India

Tomohiro Iwasaki
Department of Chemical Engineering, Osaka Prefecture University, Japan

Le Minh Duc
Faculty of Chemical Engineering, Danang University of Technology, Danang, Vietnam

Vu Quoc Trung
Faculty of Chemistry, Hanoi National University of Education, Hanoi, Vietnam